Halley

MW01482344

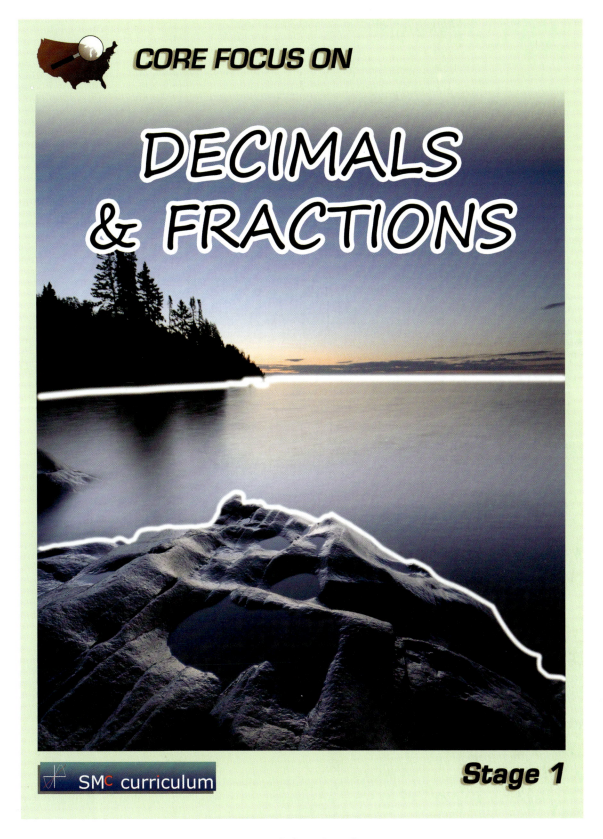

CORE FOCUS ON

DECIMALS & FRACTIONS

SMc curriculum

Stage 1

AUTHORS

SHANNON MCCAW

BETH ARMSTRONG • **MATT MCCAW** • **SARAH SCHUHL** • **MICHELLE TERRY** • **SCOTT VALWAY**

COVER PHOTOGRAPH

Lake Superior

The largest of the Great Lakes, Lake Superior borders
three US States as well as Canada to the north. It is
over 350 miles long and 150 miles wide and contains
as much water as the rest of the Great Lakes combined.
©iStockphoto.com/Andrew Parfenov

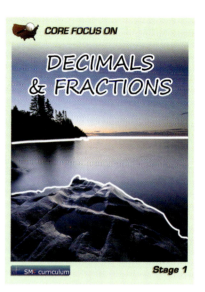

ISBN: 978-1-935033-62-2

1 2 3 4 5 6 7 8 9 10

ABOUT THE AUTHORS

SERIES AUTHOR

Shannon McCaw has been a classroom teacher in the Newberg and Parkrose School Districts. She has been trained in Professional Learning Communities, Differentiated Instruction and Critical Friends. Shannon currently works as a consultant with math teachers from over 100 districts around Oregon. Shannon's areas of expertise include the Common Core State Standards, curriculum alignment, assessment best practices and instructional strategies. She has a degree in Mathematics from George Fox University and a Masters of Arts in Secondary Math Education from Colorado College.

CONTRIBUTING AUTHORS

Beth Armstrong has been a classroom teacher in the Beaverton School District in Oregon. She has received training in Talented and Gifted Instruction. She has a Masters in Curriculum and Instruction from Washington State University.

Matt McCaw has been a classroom teacher, math/science TOSA and special education case-manager in several Oregon school districts. Matt has most recently worked as a curriculum developer and math coach for grades 6-8. He is trained in Differentiated Instruction, Professional Learning Communities, Critical Friends Groups and Understanding Poverty. Matt has a Masters of Special Education from Western Oregon University.

Sarah Schuhl has worked as an educator teaching seventh grade math through calculus and currently serves Centennial School District as the high school instructional coach and district-wide K-6 math instructional specialist. Sarah is also a Solution Tree associate and a consultant with On Target working to support teachers in the areas of math instruction and alignment to the Common Core State Standards, common assessments for all subjects and grade levels and professional learning communities. Since 2010, Sarah has been a member and chair of the National Council of Teachers of Mathematics editorial panel for their Mathematics Teacher journal. Sarah has a Masters of Science in Teaching Mathematics from Portland State University.

Michelle Terry has been a classroom teacher in the Estacada and Newberg School Districts in Oregon. Michelle has received training in Professional Learning Communities, Critical Friends, ELL Instructional Strategies, Proficiency-Based Grading and Lesson Design, Power Strategies for Effective Teaching, and Classroom Love and Logic. Michelle has an Interdisciplinary Masters from Western Oregon University. She currently teaches mathematics at Newberg High School.

Scott Valway has been a classroom teacher in the Tigard-Tualatin, Newberg and Parkrose School Districts in Oregon. Scott has been trained in Differentiated Instruction, Professional Learning Communities, Critical Friends, Discovering Algebra, Pre-Advanced Placement, Assessment Writing and Credit by Proficiency. Scott has a Masters of Science in Teaching from Oregon State University. He currently teaches math at Parkrose High School.

COMMON CORE STATE STANDARDS
Grade 6 Overview

The complete set of Common Core State Standards can be found at http://www.corestandards.org/. This book focuses on the highlighted Common Core State Standards shown below.

Ratios and Proportional Relationships

- Understand ratio concepts and use ratio reasoning to solve problems.

The Number System

- Apply and extend previous understanding of multiplication and division to divide fractions by fractions.

- Compute fluently with multi-digit numbers and find common factors and multiples.

- Apply and extend previous understandings of numbers to the system of rational numbers.

Expressions and Equations

- Apply and extend previous understandings of arithmetic to algebraic expressions.

- Reason about and solve one-variable equations and inequalities.

- Represent and analyze quantitative relationships between dependent and independent variables.

Geometry

- Solve real-world and mathematical problems involving area, surface area and volume.

Statistics and Probability

- Develop understanding of statistical variability.

- Summarize and describe distributions.

Mathematical Practices

1. Make sense of problems and persevere in solving them.

2. Reason abstractly and quantitatively.

3. Construct viable arguments and critique the reasoning of others.

4. Model with mathematics.

5. Use appropriate tools strategically.

6. Attend to precision.

7. Look for and make use of structure.

8. Look for and express regularity in repeated reasoning.

CORE FOCUS ON DECIMALS & FRACTIONS

CONTENTS IN BRIEF

How To Use Your Math Book ---- VIII

Block 1 Understanding Decimals ---- 1

Block 2 Multiplying and Dividing Decimals ---- 32

Block 3 Understanding Fractions ---- 67

Block 4 Adding and Subtracting Fractions ---- 106

Block 5 Multiplying and Dividing Fractions ---- 134

Block 6 Area and Volume ---- 167

Acknowledgements ---- 199

English/Spanish Glossary ---- 200

Selected Answers ---- 235

Index ---- 239

Problem-Solving ---- 242

Symbols ---- 243

CORE FOCUS ON DECIMALS & FRACTIONS

BLOCK 1 ~ UNDERSTANDING DECIMALS

LESSON 1.1	PLACE VALUE WITH DECIMALS	3
	EXPLORE! BASE-TEN BLOCKS	
LESSON 1.2	ROUNDING DECIMALS	8
LESSON 1.3	MEASURING IN CENTIMETERS	11
	EXPLORE! USING A METRIC RULER	
LESSON 1.4	ORDERING AND COMPARING DECIMALS	15
	EXPLORE! BATTING AVERAGES	
LESSON 1.5	ESTIMATING WITH DECIMALS	19
LESSON 1.6	ADDING AND SUBTRACTING DECIMALS	22
	EXPLORE! FIT OCCUPATIONS	
REVIEW	BLOCK 1 ~ UNDERSTANDING DECIMALS	26

BLOCK 2 ~ MULTIPLYING AND DIVIDING DECIMALS

LESSON 2.1	MULTIPLYING BY 2-DIGIT NUMBERS	34
LESSON 2.2	MULTIPLYING DECIMALS	38
	EXPLORE! SMART SHOPPER	
LESSON 2.3	DIVIDING BY 1-DIGIT NUMBERS	43
	EXPLORE! BEADED NECKLACES	
LESSON 2.4	DIVIDING BY 2-DIGIT NUMBERS	49
	EXPLORE! MAGAZINE SUBSCRIPTIONS	
LESSON 2.5	DIVIDING DECIMALS BY WHOLE NUMBERS	54
LESSON 2.6	DIVIDING DECIMALS BY DECIMALS	58
REVIEW	BLOCK 2 ~ MULTIPLYING AND DIVIDING DECIMALS	63

BLOCK 3 ~ UNDERSTANDING FRACTIONS

LESSON 3.1	GREATEST COMMON FACTOR	69
	EXPLORE! UNIVERSITY SALES	
LESSON 3.2	EQUIVALENT FRACTIONS	74
	EXPLORE! CREATING EQUIVALENT FRACTIONS	
LESSON 3.3	SIMPLIFYING FRACTIONS	79
	EXPLORE! FRACTION HOMEWORK	
LESSON 3.4	LEAST COMMON MULTIPLE	83
LESSON 3.5	ORDERING AND COMPARING FRACTIONS	87
	EXPLORE! WHICH IS LARGER?	
LESSON 3.6	MIXED NUMBERS AND IMPROPER FRACTIONS	92
	EXPLORE! CHOCOLATE CHIP COOKIES	

BLOCK 3 ~ CONTINUED

LESSON 3.7   MEASURING IN INCHES --- 96
      EXPLORE! USING A CUSTOMARY RULER

REVIEW   BLOCK 3 ~ UNDERSTANDING FRACTIONS --------------------------- 101

BLOCK 4 ~ ADDING AND SUBTRACTING FRACTIONS

LESSON 4.1   ESTIMATING SUMS AND DIFFERENCES ---------------------------- 108
LESSON 4.2   ADDING AND SUBTRACTING FRACTIONS --------------------------- 112
      EXPLORE! PIZZA PARTY!
LESSON 4.3   ADDING AND SUBTRACTING MIXED NUMBERS ----------------------- 117
      EXPLORE! MIXING PAINT
LESSON 4.4   ADDING AND SUBTRACTING BY RENAMING ------------------------- 122
LESSON 4.5   PERIMETER WITH FRACTIONS ----------------------------------- 126
REVIEW   BLOCK 4 ~ ADDING AND SUBTRACTING FRACTIONS ----------------- 130

BLOCK 5 ~ MULTIPLYING AND DIVIDING FRACTIONS

LESSON 5.1   MULTIPLYING FRACTIONS WITH MODELS -------------------------- 136
      EXPLORE! FRACTION ACTION
LESSON 5.2   MULTIPLYING FRACTIONS -------------------------------------- 139
LESSON 5.3   DIVIDING FRACTIONS WITH MODELS ----------------------------- 143
      EXPLORE! WHAT FITS?
LESSON 5.4   DIVIDING FRACTIONS --- 147
LESSON 5.5   ESTIMATING PRODUCTS AND QUOTIENTS -------------------------- 151
      EXPLORE! 4-H CLUB
LESSON 5.6   MULTIPLYING AND DIVIDING FRACTIONS AND WHOLE NUMBERS ------- 155
LESSON 5.7   MULTIPLYING AND DIVIDING MIXED NUMBERS --------------------- 159
      EXPLORE! SCRAPBOOKING
REVIEW   BLOCK 5 ~ MULTIPLYING AND DIVIDING FRACTIONS -------------- 163

BLOCK 6 ~ AREA AND VOLUME

LESSON 6.1   AREA WITH FRACTIONS -- 169
      EXPLORE! TRIANGLE AREA
LESSON 6.2   AREA AND PERIMETER WITH DECIMALS --------------------------- 174
LESSON 6.3   AREAS OF COMPOSITE FIGURES --------------------------------- 178
LESSON 6.4   NETS AND SURFACE AREAS ------------------------------------- 182
      EXPLORE! NETTING A SOLID
LESSON 6.5   VOLUME WITH FRACTIONAL DIMENSIONS -------------------------- 189
      EXPLORE! MEASURING VOLUME
REVIEW   BLOCK 6 ~ AREA AND VOLUME --------------------------------- 194

HOW TO USE YOUR MATH BOOK

Your math book has features that will help you be successful in this course. Use this guide to help you understand how to use this book.

LESSON TARGET

 Look in this box at the beginning of every lesson to know what you will be learning about in each lesson.

VOCABULARY

Each new vocabulary word is printed in red. The definition can be found with the word. You can also find the definition of the word in the glossary which is in the back of this book.

EXPLORE!

Some lessons have **EXPLORE!** activities which allow you to discover mathematical concepts. Look for these activities in the Table of Contents and in lessons next to the purple line.

EXAMPLES

Examples are useful because they remind you how to work through different types of problems. Look for the word **EXAMPLE** and the green line.

HELPFUL HINTS

Helpful hints and important things to remember can be found in green callout boxes.

BLUE BOXES

A blue box holds important information or a process that will be used in that lesson. Not every lesson has a blue box.

 This calculator icon will appear in Lessons and Exercises where a calculator is needed. Your teacher may want you to use your calculator at other times, too. If you are unsure, make sure to ask if it is the right time to use it.

EXERCISES

The **EXERCISES** are a place for you to find practice problems to determine if you understand the lesson's target. You can find selected answers in the back of this book so you can check your progress.

REVIEW

The **REVIEW** provides a set of problems for you to practice concepts you have already learned in this book. The **REVIEW** follows the **EXERCISES** in each lesson. There is also a **REVIEW** section at the end of each Block.

TIC-TAC-TOE ACTIVITIES

Each Block has a Tic-Tac-Toe board at the beginning with activities that extend beyond the Common Core State Standards. The Tic-Tac-Toe activities described on the board can be found throughout each Block in yellow boxes.

CAREER FOCUS

At the end of each Block, you will find an autobiography of an individual. Each one describes what they like about their job and how math is used in their career.

CORE FOCUS ON MATH
STAGE 1

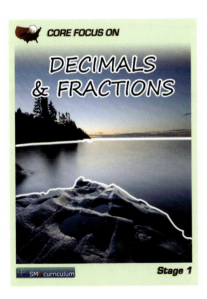

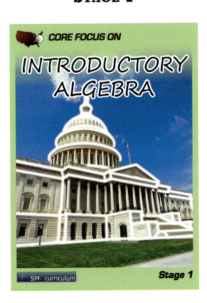

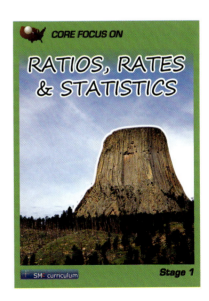

CORE FOCUS ON DECIMALS & FRACTIONS

BLOCK 1 ~ UNDERSTANDING DECIMALS

LESSON 1.1   PLACE VALUE WITH DECIMALS --- 3

        EXPLORE! BASE-TEN BLOCKS

LESSON 1.2   ROUNDING DECIMALS --- 8

LESSON 1.3   MEASURING IN CENTIMETERS -- 11

        EXPLORE! USING A METRIC RULER

LESSON 1.4   ORDERING AND COMPARING DECIMALS -- 15

        EXPLORE! BATTING AVERAGES

LESSON 1.5   ESTIMATING WITH DECIMALS -- 19

LESSON 1.6   ADDING AND SUBTRACTING DECIMALS -- 22

        EXPLORE! FIT OCCUPATIONS

REVIEW   BLOCK 1 ~ UNDERSTANDING DECIMALS --- 26

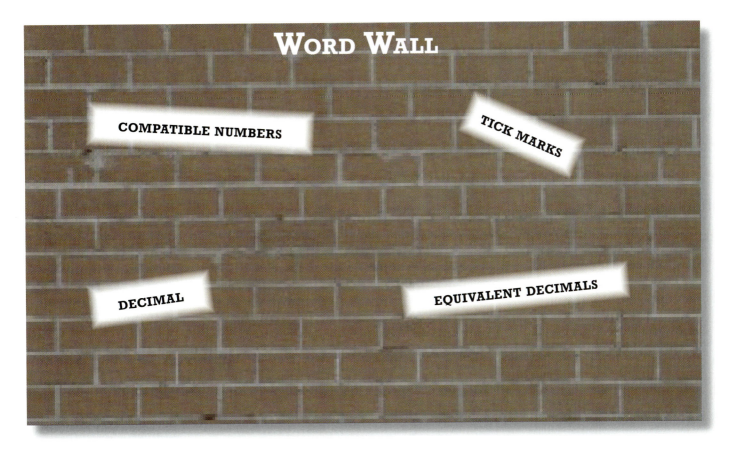

WORD WALL

COMPATIBLE NUMBERS

TICK MARKS

DECIMAL

EQUIVALENT DECIMALS

BLOCK 1 ~ UNDERSTANDING DECIMALS
TIC-TAC-TOE

PURCHASE SPREADSHEET

Add and subtract purchases using a spreadsheet.

See page 29 for details.

CHECK THE NEWSPAPER

Make a display of newspaper clippings showing examples of decimal use.

See page 7 for details.

COMPARISON SHOP

Price foods you eat at two different stores.

See page 7 for details.

ESTIMATION RAP

Write a rap song about situations where estimating is acceptable and where estimating would cause disasters.

See page 21 for details.

MEAL DEALS

Create a menu with prices and meal deals.

See page 7 for details.

CHECKBOOK REGISTRY

You have $300 to spend on gifts for family and friends. Record purchases in your checkbook registry.

See page 29 for details.

METRIC MADNESS

Explore the metric system. Make a table showing the different units of measurement in the metric system.

See page 25 for details.

PLACE VALUE STORY

Write a picture book about place value. Use decimals as characters.

See page 18 for details.

DECIMAL DASH

Create a board game using addition and subtraction of decimals.

See page 25 for details.

PLACE VALUE WITH DECIMALS

 Identify place value of decimals to the thousandths.

The fruit stand charges $2.75 for a pound of dried fruit. Trish bought 4.1 pounds of dried fruit. She paid $11.55 for the fruit.

2.75 ←— decimal

↑
decimal point

All numbers in the situation above are decimals. A **decimal** is any base ten number written with a decimal point. Decimals are based on the number ten.

FOUR DIFFERENT WAYS TO MODEL DECIMALS

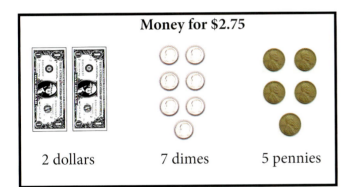

Money for $2.75

2 dollars   7 dimes   5 pennies

Place-Value Chart showing 2.75

1000	100	10	1	0.1	0.01	0.001
Thousands	Hundreds	Tens	Ones	Tenths	Hundredths	Thousandths
			2	7	5	

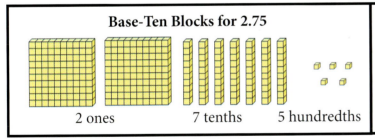

Base-Ten Blocks for 2.75

2 ones   7 tenths   5 hundredths

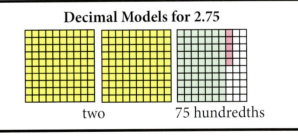

Decimal Models for 2.75

two   75 hundredths

The value and position of each digit in a decimal determine how much the decimal is worth. The decimal point separates the whole number from the part that is less than one.

1000	100	10	1	0.1	0.01	0.001
Thousands	Hundreds	Tens	Ones	Tenths	Hundredths	Thousandths
		1	3	5	6	2

Whole Number   Less Than One

What is it worth?
The 1 represents one ten.
The 3 represents three ones.
The 5 represents five tenths.
The 6 represents six hundredths.
The 2 represents two thousandths.

To write decimals in word form, separate the whole number from the part that is less than one with the word "and." The decimal point is read as "and."

EXAMPLE 1

Write **431.25** in word form.

SOLUTION

Fill in the place-value chart for the number.

1000	100	10	1	0.1	0.01	0.001
Thousands	Hundreds	Tens	Ones	Tenths	Hundredths	Thousandths
	4	3	1 •	2	5	

Two tenths is the same as twenty hundredths.

The last digit is in the hundredths place.

Four hundred thirty-one   and   twenty-five hundredths

EXAMPLE 2

Write **45.703** in word form.

SOLUTION

Fill in the place-value chart for the number.

1000	100	10	1	0.1	0.01	0.001
Thousands	Hundreds	Tens	Ones	Tenths	Hundredths	Thousandths
		4	5 •	7	0	3

The last digit is in the thousandths place.

Forty-five   and   seven hundred three thousandths

When there is no whole number preceding the decimal point, only the part that is less than one is read.

EXAMPLE 3

Write **0.6** in word form.

SOLUTION

Fill in the place-value chart for the number.

1000	100	10	1	0.1	0.01	0.001
Thousands	Hundreds	Tens	Ones	Tenths	Hundredths	Thousandths
			0 •	6		

The last digit is in the tenths place.

Six tenths

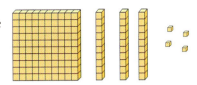

Step 1: Set out base-ten blocks for the decimal 1.34 and fill in a place-value chart. Write this number in word form. Which number is in the tenths place?

Step 2: Set out base-ten blocks for the decimal 4.51 and fill in a place-value chart. Write this number in word form. Which number is in the hundredths place?

Step 3: Set out base-ten blocks for the decimal 2.97 and fill in a place-value chart. Write this number in word form. Which number is in the hundredths place?

Step 4: Set out base-ten blocks for the decimal 5.09 and fill in a place-value chart. Write this number in word form. Which number is in the tenths place?

Step 5: Set out base-ten blocks for the decimal 3.02 and fill in a place-value chart. Write this number in word form. Which number is in the hundredths place?

Step 6: Set out base-ten blocks for the decimal 2.53 and fill in a place-value chart. Which number is in the tenths place?

Step 7: Sketch a model of 2.46 as base-ten blocks on a piece of paper.

Sketching shortcuts:

ones	tenths	hundredths

Step 8: Fill in a place-value chart with the decimal that matches your drawing.

Step 9: Write the word form for your decimal under the place-value chart.

Step 10: Repeat **Steps 7-9** for at least three decimals using ones, tenths and hundredths.

EXAMPLE 4

SOLUTION

Write the decimal that represents two and sixty-three hundredths.

Write the whole number.	two = 2
Insert the decimal point for 'and.'	two and = 2.
Add the part that is less than one.	sixty-three hundredths = 2.63

two and sixty-three hundredths = 2.63

EXERCISES

Write the decimal that each base-ten block group represents.

1.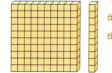

2.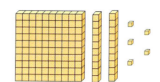

3.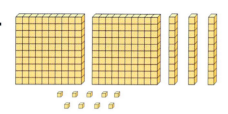

4.

5.

6.

7.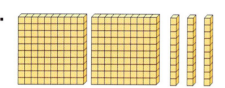

8.

9. In which place is the digit 9 in each decimal?

    a. 42.709   **b.** 9.231   **c.** 14.902   **d.** 81.493

Write the value of the underlined digit in each decimal.

10. 2.1̱83

11. 31.20̱8

12. 7.26̱4

13. 43̱.56

Write a decimal to match each word form.

14. sixteen and twenty-three hundredths

15. ninety-nine and two tenths

16. three and four hundred eleven thousandths

17. fifty-two and six hundredths

18. seventy-eight and four thousandths

19. two hundred twenty-three thousandths

Write each decimal in word form.

20. 4.9

21. 35.87

22. 11.04

23. 0.549

24. 4.008

25. 68.4

Tic-Tac-Toe ~ Comparison Shop

Step 1: List 10 food items found in your home (include brand name and container size.)

Step 2: Choose two different local grocery stores. "Shop" the two stores for the food items on your list by pricing the same sized packages of the items on your list. You will not actually be buying the food items. See example of foods below.

Food Item	Store #1	Store #2
Circle Oat Cereal 12 *oz* box	$3.98	$2.50
Store Brand Milk 1 gallon	$2.49	$2.69
Banana per pound	$0.54	$0.45

Step 3: Total the costs at each store. Decide which store had the best prices on the food items on your list.

Tic-Tac-Toe ~ Meal Deals

Create a menu for a restaurant to open for students complete with fair prices (including decimal amounts) on at least 20 different food items.

On the menu group several individual items together as a "Meal Deal" that students can buy instead of purchasing each item separately. Add prices of individual food items that are grouped together. Then round the total to a more appealing "Meal Deal" price.

Design your menu. You can design it on a computer or draw it and neatly print. Give a title to your restaurant. The menu should represent your restaurant (i.e., If you design a breakfast menu, the menu could have pictures of breakfast food items you serve). The menu should be easy to read and understand.

Tic-Tac-Toe ~ Check the Newspaper

You say "half-hour" not "0.5 hour" (point five of an hour). You say you spent $1.50, not $1\frac{1}{2}$ dollars. There are many instances that require you to write or speak using decimals. Look through several newspapers. Cut out clippings showing decimal use. Make a display showing examples where decimals are used.

Choose four examples. Explain why the chosen method of writing decimals works for each example.

ROUNDING DECIMALS

 Round decimals to the nearest one, tenth, hundredth or thousandth.

Many products are labeled with both customary and metric measurements. For example, a half gallon of milk is also labeled 1.89 liters. If you want to know about how many liters equal a half gallon, you can round the measurement. Look at the digits that follow the one you want to round to (in this case it would make sense to round to the ones place). Rounding 1.89 liters to the nearest liter is 2 liters. Two liters is approximately equal to a half gallon.

ROUNDING DECIMALS

1. Underline the digit to which you will round.
2. Look at the digit to the right of the underlined digit.
   - If the digit is 4 or less, the underlined numeral stays the same.
   - If the digit is 5 or greater, add one to the underlined digit.
3. Rewrite the decimal. Stop after writing the rounded digit.

EXAMPLE 1

Round 0.59 to the nearest tenth.

SOLUTION

Underline the number in the tenths place.   0.5̲9

Look at the digit to the right of the underlined digit.   It is 9.

If the digit is 5 or greater, add one to the underlined digit.   5 + 1 = 6

0.59 rounded to the nearest tenth is 0.6.

EXAMPLE 2

A newborn bird weighed 2.84 ounces. How much did the bird weigh to the nearest ounce?

SOLUTION

Underline the digit you are rounding to.   2̲.84

Look at the digit to the right of the underlined digit.   It is 8.

If the digit is 5 or greater, add one to the underlined digit.   2 + 1 = 3

The baby bird weighed almost 3 ounces because 2.84 rounded to the nearest one is 3.

8 *Lesson 1.2 ~ Rounding Decimals*

EXAMPLE 3

Terrell worked 15.25 hours last week. He is paid for each full hour he works. When he fills in his time card, he must fill in the hours to the nearest hour. What should he write for last week?

SOLUTION

Terrell needs to round to the nearest one.   15.25

Look at the digit to the right of the underlined digit.   It is a 2.

If the digit is 4 or less, the underlined digit stays the same.

15.25 rounded to the nearest one is 15.

Terrell needs to write 15 hours on his time card.

EXAMPLE 4

A gas station advertised the price of unleaded gas as $3.999 per gallon. Oksana said she paid $3.99 per gallon. Was she correct?

SOLUTION

Oksana needs to round to the nearest hundredth (penny).   3.999

Look at the digit to the right of the underlined digit.   It is 9.

If the digit is 5 or greater, add a one to the underlined digit.   9 + 1 = 10

Because ten is not a single digit, continue rounding digits to the left until you do not get a ten. In this case, $3.999 rounded to the nearest penny (hundredth) becomes $4.00.

Oksana is not correct. She actually paid closer to $4.00 per gallon of gas than $3.99.

EXERCISES

Round each number to the nearest one.

1. 4.3

2. 2.523

3. 6.89

4. 23.09

5. 17.99

6. 9.99

7. Joshua spent 1.75 hours texting his friends last week. To the nearest hour, how many hours did he spend texting last week?

8. A meter equals 39.37 inches. About how many inches equal one meter?

Round each number to the nearest tenth.

9. 34.91

10. 43.56

11. 71.25

12. 3.57

13. 3.908

14. 4.96

15. Maria wants to buy sheet music for $2.36. How much money should she take to buy it, to the nearest dime?

16. Aleah downloaded two music videos. They cost a total of $3.88. How much will she spend, to the nearest dime?

Round each number to the nearest hundredth.

17. 45.205

18. 1.632

19. 321.2371

20. 6.228

21. 3.904

22. 1.0962

Round each number to the nearest thousandth.

23. 7.9087

24. 50.1011

25. 201.10895

26. 1.7999

27. 5.000647

28. 11.3192

REVIEW

Write a decimal that matches each base ten block group.

29.
30.
31.

Write each decimal in word form.

32. 2.1

33. 70.07

34. 3.012

Find the value of the underlined digit in each decimal.

35. 4.7<u>8</u>2

36. 51.<u>2</u>4

37. 17.82<u>1</u>

MEASURING IN CENTIMETERS

 Measure and draw line segments using centimeters.

The **metric system** is used in a majority of countries around the world. The metric system is a decimal system of measurement. The basic unit of length in the metric system is the meter. Other metric units, such as millimeters, centimeters and kilometers, are based on the meter.

The United States is one of only a few countries that use the customary system as its primary system of measurement. Customary units of length include inches, feet, yards and miles.

There are many reasons you should learn about the metric system. When you purchase items made in other countries the measurements are often in metric units. You may travel in other countries where you need to read signs in metric units. Can you think of other situations where you might see measurements in metric units?

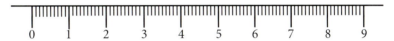

The metric ruler is divided up using **tick marks**. Tick marks are equally divided spaces marked with a small line. Each whole number on a metric ruler represents a centimeter (*cm*). The small tick marks between each centimeter represent millimeters (*mm*). There are 10 millimeters in 1 centimeter. Each millimeter is one-tenth of a centimeter.

EXAMPLE 1

Find the length of the line to the nearest tenth of a centimeter.

SOLUTION

Line up the 0 mark on the ruler with the left edge of the line. Identify the tick mark that represents the length of the line.

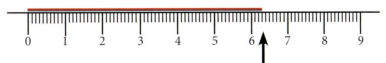

Each small tick mark represents 0.1 centimeters. Count the number of tick marks after the last whole number to find the number of tenths.

6	+	0.3	=	6.3 *cm*
WHOLE NUMBER	+	TENTHS	=	TOTAL LENGTH

The length of the line is 6.3 *cm*.

Step 1: Use your ruler to measure the length of each object below to the nearest tenth of a centimeter. Record your answers.

Step 2: Sometimes measurements are approximated to the nearest half centimeter.
   a. Round each of the measurements from **Step 1** to the nearest half centimeter.
   b. What two decimal numbers can the measurements end with if they are rounded to the nearest half centimeter?

Step 3: Draw a line that fits each description.
   a. exactly 3.6 centimeters long   **d.** exactly 7 centimeters long
   b. exactly 0.8 centimeters long   **e.** approximately 3.5 centimeters long
   c. approximately 7 centimeters long   **f.** approximately 1.5 centimeters long

Step 4: Are the lines from **part c and d** in **Step 3** the exact same length? Do they have to be? Why or why not?

Step 5: Estimate how long your pencil is in centimeters. Record your estimate. Measure your pencil to see how long it is. How far off was your estimate to the nearest tenth of a centimeter?

Step 6: Estimate the length (in centimeters) of two other objects in your classroom. Measure the objects to the nearest half centimeter. Approximately how far off were your estimates?

EXAMPLE 2   **Find the length of the line to the nearest half centimeter.**

SOLUTION   Line up the 0 mark on the ruler with the left edge of the line. Identify the tick mark that represents the length of the line.

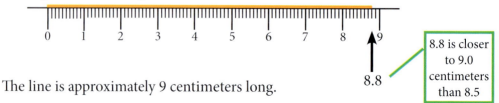

8.8 is closer to 9.0 centimeters than 8.5

The line is approximately 9 centimeters long.

EXERCISES

1. How many spaces is each centimeter divided into by the tick marks?

2. Round 3.2 to the nearest half centimeter.

3. Round 4.8 to the nearest half centimeter.

4. A line measured 3 centimeters and 4 millimeters. Write this measurement as a decimal with centimeter units.

5. A pen measured 14 centimeters and 1 millimeter. Write this measurement as a decimal with centimeter units.

Find the length of each line to the nearest tenth of a centimeter.

6. ────────────────────────────────

7. ──────────────────────────

8. ──────────────────────────────────────

9. ───────────────────────────────────

10. ───

11. ───────────

12. Use the pocket watch to the right.
   a. Estimate the width of the pocket watch to the nearest tenth of a centimeter.
   b. Measure the width of the pocket watch to the nearest tenth of a centimeter.
   c. How far off was your estimate, to the nearest tenth of a centimeter?

13. Use the bobby pin below.

   a. Estimate the length of the bobby pin to the nearest tenth of a centimeter.
   b. Measure the length of the bobby pin to the nearest tenth of a centimeter.
   c. How far off was your estimate, to the nearest tenth of a centimeter?

Find the length of each line to the nearest half centimeter.

14. ─────────────────────────

15. ──────────────────────

16. ━━━━━━━━━━━━━━━━━━━━

17. ━━━━━━━━━━━━━━━━━━━━━━━

18. ━━━━━━━

19. Use the key.

    a. Estimate the length of the key to the nearest half centimeter.
    b. Measure the length of the key to the nearest half centimeter.
    c. How far off was your estimate, to the nearest half centimeter?

20. Use the lemon.
    a. Estimate the length of the lemon to the nearest half centimeter.
    b. Measure the length of the lemon to the nearest half centimeter.
    c. How far off was your estimate, to the nearest half centimeter?

Draw a line with each given length.

21. 3.4 *cm*   **22.** 4.2 *cm*   **23.** 5.7 *cm*

24. 9.5 *cm*   **25.** 0.9 *cm*   **26.** 8 *cm*

27. Shari needed a length of ribbon that measured at least 34.8 centimeters long. She rounded her measurement to the nearest centimeter. How long is the piece of ribbon?

28. Fernando and Maria measured the black line below. Fernando said it measured 5 centimeters long. Maria said the line measured 4.7 centimeters long. ━━━━━━━
    a. Whose measurement was more accurate?
    b. What might the other person have done to get their measurement?

REVIEW

Round each number to the nearest tenth.

29. 15.74   **30.** 278.409   **31.** 321.35

Round each number to the nearest hundredth.

32. 1.408   **33.** 215.172   **34.** 18.976

ORDERING AND COMPARING DECIMALS

 Order and compare decimals to find the smallest or largest decimal.

The ladies' short program for ice skating at the 2006 Olympic Winter Games in Torino ended with the following results:

Ice Skater	Total Segment Score
Irina Slutskaya	66.7
Shizuka Arakawa	66.02
Sasha Cohen	66.73

Source: http://www.torino2006.org

Comparing decimals is similar to comparing whole numbers.

Decimals that do not have the same number of digits after the decimal point can be compared by inserting zeros to hold place value. Decimals that name the same amount are called **equivalent decimals**.

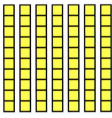 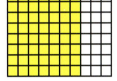

0.7 (seven tenths) = 0.70 (seventy hundredths)

EXAMPLE 1

Use place value to list the scores of Irina Slutskaya (66.7), Shizuka Arakawa (66.02) and Sasha Cohen (66.73) from greatest to least.

SOLUTION

Line up the decimal points.

66.70
66.02
66.73

> Irina Slutskaya's score of 66.7 = 66.70

> Insert a zero to hold place value.

The tenths place is different.

Starting from the left and moving right, compare the digits until there is a digit that differs.

Compare the digits in the tenths place.

0 < 7 so 66.02 is the smallest.

Compare the digits in the hundredths place for the two remaining numbers.

66.70
66.73
0 < 3 so 66.70 < 66.73

Sasha had the highest score (66.73). Irina had the next highest score (66.7). Shizuka had the lowest score (66.02) of the three figure skaters.

EXAMPLE 2

Use a number line to compare 7.4 and 7.38.

SOLUTION

Decimals on a number line get larger as you move from left to right.

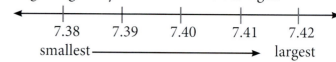

7.38   7.39   7.40   7.41   7.42
smallest ——————————→ largest

Put the decimals on a number line.

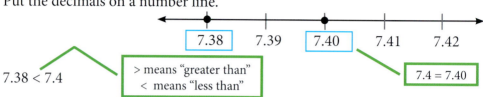

7.38   7.39   7.40   7.41   7.42

7.38 < 7.4

> means "greater than"
< means "less than"

7.4 = 7.40

EXAMPLE 3

Put the following numbers in order from least to greatest.
12, 12.11, 12.01, 12.113

SOLUTION

Line up the decimal points. Insert zeros so each number has the same number of places after the decimal point.

12.000
12.110
12.010
12.113

Use place value to compare digits.

The order from least to greatest is: 12, 12.01, 12.11, 12.113

EXPLORE!

BATTING AVERAGES

PLAYER	BATTING AVERAGE
Lin	0.15
Jacie	0.187
Sorrell	0.300
Libby	0.143
Octavia	0.24
Jaelynn	0.14
Maria	0.2
Leslie	0.198
Julia	0.31
Thora	0.185
Kate	0.141
Brynn	0.240
Sarah	0.252
Isabelle	0.3

The Jackson Middle School fastpitch coach kept track of the players' batting averages. At the end of the season, the batting averages were posted (see the table on the left). Batting averages are usually listed to the thousandths place. When the averages were posted, some were only listed to the tenths or hundredths place. Help the players figure out their batting averages. (Note: The larger the decimal, the better the batting average.)

Step 1: Who had the best batting average? How do you know?

Step 2: Who had the worst batting average? How do you know?

Step 3: Use >, < and = to compare batting averages.
    a. Jacie and Thora   **d.** Lin and Julia
    b. Isabelle and Sorrell   **e.** Kate and Jaelynn
    c. Octavia and Brynn   **f.** Jaelynn and Libby

16 *Lesson 1.4 ~ Ordering and Comparing Decimals*

Step 4: List each group from least to greatest according to their batting averages.
    a. Maria, Octavia, Leslie, Kate
    b. Lin, Jacie, Isabelle, Sarah, Jaelynn
    c. Brynn, Thora, Julia, Maria, Libby

Step 5: Put the entire team's batting averages in order from least to greatest.

EXERCISES

Replace each ⬤ with <, > or = to make a true sentence.

1. 3.1 ⬤ 3.2

2. 7.03 ⬤ 7.3

3. 5.751 ⬤ 5.75

4. 6.5 ⬤ 6.50

5. 42.9 ⬤ 42.19

6. 4.567 ⬤ 4.678

7. 2.140 ⬤ 2.104

8. 32.7 ⬤ 32.70

9. 1.11 ⬤ 1.111

Put each set of numbers in order from least to greatest.

10. 4.45, 4.4, 4.44, 4.42

11. 17.801, 17.81, 17.8, 17.851

12. 5.9, 5.99, 5.09, 5.999

Choose the best answer for each question.

13. Which number is between 6.77 and 6.97?
    A. 6.7
    B. 6.98
    C. 6.9
    D. 6.07

14. Which number is larger than 2.421?
    A. 2.42
    B. 2.425
    C. 2.4
    D. 2.411

15. Which number is smaller than eight and eight hundred four thousandths?
    A. 8.9   **B.** 8.805   **C.** 8.84   **D.** 8.8

Many states have a minimum wage law. The least amount an employer can pay an employee is this wage per hour. Use the 2007 Minimum Wage chart to answer each question.

16. Which state has the highest minimum wage?

17. Which state has the lowest minimum wage?

18. Put the six states in order from lowest minimum wage to highest minimum wage.

State	Minimum Wage
Alaska	$7.15
Oregon	$7.80
Washington	$7.93
California	$7.50
Idaho	$5.85
Nevada	$6.33

http://www.dol.gov.miniwage/America.html

Nine different runners' times for the 400-meter dash are displayed in the table. Use the information in the table to complete the following problems.

400-meter dash time (in seconds)
50.5
49.09
49.45
49.76
49.25
51.5
48.9
50.98
49.61

19. _____ is the fastest time.

20. _____ and _____ are slower than 50.55 seconds.

21. _____ is the slowest time.

22. _____ and _____ are faster than 49.2 seconds.

23. _____ and _____ are between 50.45 and 51.05 seconds.

24. _____ is equal to 49.760 seconds.

25. Put the times in order from least to greatest (fastest to slowest).

26. Ethan drank 2.2 liters of water on Friday and 2.25 liters of water on Saturday. On which day did he drink less water?

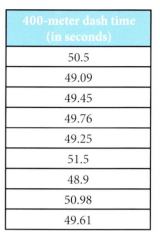

27. The temperature on Monday was 68.75° F. On Tuesday, it was 68.8° F. Wednesday's temperature was 68.08° F.
   a. Which day was the hottest?
   b. Which day was the coolest?
   c. Put the three temperatures in order from least to greatest.

REVIEW

Round each number to the underlined digit.

28. <u>4</u>.507

29. 32.6<u>3</u>4

30. 7.<u>9</u>86

Write each number in word form as a decimal.

31. Two and seventeen hundredths

32. Fifty and two thousandths

33. Six and four tenths

TIC-TAC-TOE ~ PLACE VALUE STORY

Fiction picture books contain short stories that have a problem and a solution. Write your own fiction story about place value with decimals as your characters. Their problem is that they can not figure out which one of them is the largest. Make sure your story includes a solution to the problem. A complete fiction picture book will include illustrations, a cover and a title.

ESTIMATING WITH DECIMALS

 Estimate sums, differences, products or quotients of expressions involving decimals.

Washington County could publish the following information:

There are 10,500 students in the county. Washington County has 28 public schools.

Which of the statements is likely an estimate?
Which is an exact statement?

There are situations where using an estimate makes sense. The number of students in Washington County Schools probably changes every day because students move in and out of the county. The number of schools does not change often, so this is an exact statement.

Estimates are sometimes given instead of exact answers. It is confusing to say that the average American family has 2.4 children. Instead, it is often said that the average American family has about two children. The most common method for estimating decimal expressions is to round to the nearest whole number before calculating.

EXAMPLE 1

The Western Meadowlark weighs between 3.18 and 5.3 ounces. Estimate the difference between the heaviest weight and the lightest weight given.

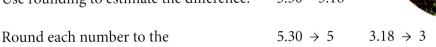

SOLUTION

Use rounding to estimate the difference.　　$5.30 - 3.18$

Round each number to the nearest whole number.　　$5.30 \rightarrow 5$　　$3.18 \rightarrow 3$

Subtract.　　$5 - 3 = 2$　so　$5.3 - 3.18 \approx 2$

The difference between the heaviest and lightest weight for the Western Meadowlark is about 2 ounces.

EXAMPLE 2

Apples cost $1.97 per pound. Ieysha buys a bag of apples that weighs 4.7 pounds. About how much money did Ieysha need to purchase the apples?

SOLUTION

Use rounding to estimate the product.　　$\$1.97 \times 4.7$

Round each number to the nearest whole number.　　$\$1.97 \rightarrow \$2$　　$4.7 \rightarrow 5$

Multiply.　　$\$2 \times 5 = \$10$ so $\$1.97 \times 4.7 \approx \$10$

Ieysha needed about $10.00 to purchase the apples.

Compatible numbers can be used when estimating quotients. **Compatible numbers** are numbers that are easy to mentally compute. Round the divisor to the nearest whole number. Change the dividend to the nearest multiple of the new divisor. This makes the two numbers compatible, or easy to compute mentally.

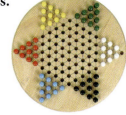

EXAMPLE 3

Maris spent $121.28 on board games for her nieces and nephews. The board games she bought cost around $19.95 each. Approximately how many board games did she buy?

SOLUTION

Use compatible numbers to estimate. $121.28 ÷ $19.95

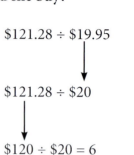

Round the divisor to the nearest whole number. $121.28 ÷ $20

Change the dividend to the nearest multiple of the new divisor. $120 ÷ $20 = 6

Maris bought about six board games.

EXERCISES

Use rounding to the nearest whole number to estimate each sum or difference.

1. 4.6 + 2.9

2. 5.678 + 1.231

3. 7.9 + 5.21

4. 2.12 − 1.04

5. 9.61 − 3.3

6. 12.99 − 4.6

7. 8.73 + 8.29 + 8.9

8. 11.051 + 12.523 + 2.423

9. 15.4 + 4.8 + 2.4

10. Zade is the cashier for his family's garage sale. A lady wants to buy a bag of clothes for $5.75 plus a few books for $5.40. Approximately how much money will she owe Zade?

11. Xavier took $33.25 to the store. He bought a CD that cost $17.85. About how much money did he have left?

Use rounding to the nearest whole number to estimate each product.

12. 6.22 × 3.54

13. 3.98 × 4.123

14. 9.84 × 4.89

15. 9.04 × 3.13

16. 19.941 × 4.784

17. 36.05 × 6.322

18. Jessica buys 3.4 pounds of bulk candy. It costs $1.78 per pound. Approximately how much will the candy cost Jessica?

19. Jorgé earned $10.60 per hour for 19.25 hours of work. About how much did he earn?

Use compatible numbers to estimate each quotient.

20. 123.4 ÷ 24.95

21. 33.567 ÷ 7.05

22. 93.3 ÷ 9.05

23. 30.901 ÷ 11.167

24. 51.33 ÷ 2.211

25. 21.01 ÷ 19.275

26. Stefani orders 8.8 yards of fabric. It takes 2.5 yards of fabric for each of her projects. Estimate how many projects Stefani can sew with the fabric she ordered.

27. Mya went to the bookstore to buy books for herself. She had $59.15 to spend. Each book she bought cost $5.95. About how many books did she buy?

REVIEW

Replace each ⬤ with <, > or = to make a true sentence.

28. 2.8 ⬤ 2.85

29. 7.010 ⬤ 7.01

30. 6.755 ⬤ 6.76

Put each group of numbers in order from least to greatest.

31. 32.4, 32.43, 32.34, 32.48

32. 17.9, 18, 17.09, 18.9

33. 11.02, 11.22, 11.022, 11.2

Draw a line with each given length.

34. 3.2 *cm*

35. 1.7 *cm*

36. 2.8 *cm*

37. 0.5 *cm*

38. 2.1 *cm*

39. 1.4 *cm*

TIC-TAC-TOE ~ ESTIMATION RAP

There are times where estimation is accepted and an exact answer is not needed. However, if you always estimate, problems might arise. For example, if you estimate by rounding to the nearest dollar for something that cost $2.15, you may only bring $2.00 with you. This would not be enough money to buy the item.

Make a list of situations where it is acceptable to estimate. Make another list of times where you should not estimate. Create a rap song using the lists.

ADDING AND SUBTRACTING DECIMALS

Find sums or differences of expressions involving decimals.

Laura walked one mile (1,609.344 meters) on Friday. She walked three-quarters of a mile (1,207.008 meters) on Saturday. She wants to keep track of her total meters walked. She will need to add these decimals together.

ADDING OR SUBTRACTING DECIMALS

1. Line up the decimal points.
2. Insert zeros so each decimal has the same amount of places after the decimal point.
3. Add or subtract.
4. Move the decimal point down into the answer in its same position.

EXAMPLE 1

Laura walked 1,609.344 meters on Friday and 1,207.008 meters on Saturday. How many meters did she walk altogether?

SOLUTION

Write the problem.	$1609.344 + 1207.008$
Line up the decimal points.	$\begin{array}{r} 1609.344 \\ +\ 1207.008 \\ \hline \end{array}$
Add.	$\begin{array}{r} \overset{1}{\ }\ \overset{1}{\ } \\ 1609.344 \\ +\ 1207.008 \\ \hline 2816.352 \end{array}$

Laura walked 2,816.352 meters altogether.

EXAMPLE 2

Laura walked 1,628.3 meters on Sunday and 1,207.25 meters on Monday. How many more meters did Laura walk on Sunday than on Monday?

SOLUTION

Write the problem.	$1628.3 - 1207.25$
Line up the decimal points.	$\begin{array}{r} 1628.3 \\ -\ 1207.25 \\ \hline \end{array}$
Insert zeros.	$\begin{array}{r} 1628.30 \\ -\ 1207.25 \\ \hline \end{array}$
Subtract.	$\begin{array}{r} {}^{2}\ {}^{10} \\ 1628.\cancel{30} \\ -\ 1207.25 \\ \hline 421.05 \end{array}$

> Zeros can be added at the end of a decimal without changing the value of the number.

On Sunday, Laura walked 421.05 meters more than on Monday.

Many Americans are not exercising enough according to a study published in 2007 by the American Council of Exercise. The table below shows the results from a study documenting how many steps (and the conversion of steps to miles) were taken by people in different occupations.

The average total distance in miles is shown in the last column. For example, the secretaries in this study walked, on average, 1.7 ± 0.66 miles each day. This means:

Shortest Distance	Longest Distance
1.7 − 0.66	1.7 + 0.66
6 10	1
1.7̶0̶	1.70
− 0.66	+ 0.66
1.04	2.36

Table 1. Average steps and distance walked by people in different occupations over the course of an average working day.		
Occupation	Total Steps	Total Distance (mi)
Secretaries	4,327 ± 1,671	1.7 ± 0.66
Teachers	4,726 ± 1,832	1.9 ± 0.73
Lawyers	5,062 ± 1,837	2.0 ± 0.73
Police officers	5,336 ± 1,767	2.1 ± 0.70
Nurses	8,648 ± 2,461	3.4 ± 0.98[a]
Construction workers	9,646 ± 2,719	3.8 ± 1.08[a]
Factory workers	9,892 ± 2,496	3.9 ± 0.99[a]
Restaurant servers	10,087 ± 2,908	4.0 ± 1.15[a]
Custodians	12,991 ± 4,902	5.2 ± 1.94[a,b]
Mail carriers	18,904 ± 5,624	7.5 ± 2.23[a,b,c]

[a]Significantly different than secretaries, teachers, lawyers and police officers ($p < 0.05$).

[b]Significantly different than nurses, construction workers, factory workers and restaurant servers ($p < 0.05$).

[c]Significantly different than all other occupations ($p < 0.05$).

Source: http://www.acefitness.org

Use the table to answer the following questions.

Step 1: What was the longest average total distance walked each day by:
　　a. nurses?
　　b. restaurant servers?
　　c. mail carriers?

Step 2: What was the shortest average total distance walked each day by:
　　a. construction workers?
　　b. lawyers?
　　c. mail carriers?

Step 3: Use the first number of each expression in the total distance column to determine how much further on average _____ walked each day than _____ .
　　a. mail carriers, secretaries?
　　b. construction workers, teachers?
　　c. restaurant servers, police officers?

Example: custodians and nurses
Custodians' 1ˢᵗ number: 5.2 ± 1.94
Nurses' 1ˢᵗ number: 3.4 ± 0.98

4 12
5.2̶
− 3.4
1.8

Custodians walked, on average, 1.8 miles more than nurses.

EXERCISES

Find each sum.

1. 2.1 + 3.4

2. 4.32 + 5.29

3. 3.786 + 9.42

4. 4.607 + 3.4

5. 1.325 + 5.78

6. 53.999 + 32.187

7. Monica spent $15.17 on a pair of pants and $4.96 on a pair of socks. How much did she spend altogether?

Find each difference.

8. 7.2 – 2.5

9. 4.31 – 1.75

10. 8.241 – 6.456

11. 12.1 – 9.24

12. 6.087 – 3.43

13. 15.55 – 11.901

14. Chan filled his car with 12.85 gallons of gas one week. The next week he filled his car with 9.08 gallons. How many more gallons of gas did he put in his car the first week than the second week?

The table at the right shows the monthly rainfall in Ocean View for 2005. Use the table to answer each question.

Month	Rainfall (inches)
January	8.07
February	2.85
March	6.22
April	5.02
May	5.86
June	3.31
July	1
August	0.03
September	2.72
October	5.59
November	8.44
December	15.58

15. How many inches of rain fell in Ocean View in September and October altogether?

16. How many more inches of rainfall did Ocean View have in November than in October?

17. How many more inches of rainfall did Ocean View have in December than in March?

18. How many inches of rainfall did Ocean View have in May and June altogether?

19. How many total inches of rainfall did Ocean View accumulate in February, March and April?

20. How many total inches of rainfall did Ocean View have in the last two months of the year?

21. What was the total rainfall in Ocean View during the first two months of the year?

22. Describe the process of adding two decimals.

23. Nachelle had $271.74 in her checking account. She wrote two checks. One was for $52.49 and the other was for $14.88. How much money did she have left in her checking account?

REVIEW

24. Write a decimal that would be between 1.5 and 2.

25. Write a decimal that is bigger than 3.4 but smaller than 3.5.

26. Write a decimal that is smaller than 5.056 but bigger than 5.05.

27. Round 4.678 to the nearest hundredth.

28. Round 23.969 to the nearest tenth.

29. Round 4.3809 to the nearest thousandth.

30. A line measured 8 centimeters and 2 millimeters. Write this measurement as a decimal with centimeter units.

31. A computer measured 36 centimeters and 8 millimeters. Write this measurement as a decimal with centimeter units.

TIC-TAC-TOE ~ DECIMAL DASH

Create a "Decimal Dash" board game with a path marked off in boxes with at least 30 boxes. Each box is given a decimal value of 0.2 to 1.2 which must be used more than once.

Use two different colored dice to play the game. The dots on each die represent tenths (1 dot = 0.1, 2 dots = 0.2)

Game rules:

Step 1: Choose one colored die for addition and one colored die for subtraction.
(i.e., Red die = addition, Blue die = subtraction)

Step 2: Roll both dice.
   EXAMPLE 1: Roll both dice: Red rolls as 0.6; Blue rolls as 0.4.
      0.6 > 0.4 so Red die is the larger decimal. You need to add.
      0.6 + 0.4 = 1.0
      Move forward to the next 1.0 space.

   EXAMPLE 2: Roll both dice. Red rolls as 0.2; Blue rolls as 0.5.
      0.2 < 0.5 so Blue die is the larger decimal. You need to subtract.
      0.5 – 0.2 = 0.3
      Move backward to the closest 0.3 space.

Step 3: The first one to end (your choice as to what location is) wins.

TIC-TAC-TOE ~ METRIC MADNESS

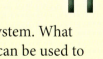

Investigate the metric system using the internet or class resources.

Create a table showing the different units of measurement in the metric system. What metric units can be used to measure length or height? What metric units can be used to measure mass? What metric units can be used to measure liquids?

Write a paragraph about the metric system. In what ways do you find the metric system different than the customary system (inches, feet, etc.) used in the United States? Which system do you prefer and why?

Identify place value of decimals to the thousandths.
Round decimals to the nearest one, tenth, hundredth or thousandth.
Measure and draw line segments using centimeters.
Order and compare decimals to find the smallest or largest decimal.
Estimate sums, differences, products or quotients of expressions involving decimals.
Find sums or differences of expressions involving decimals.

Lesson 1.1 ~ Place Value with Decimals

Write the decimal that matches each base-ten block model.

1.

2.

3.

In which place is the digit 5 in each decimal?

4. 15.208

5. 3.521

6. 1.005

Write the decimal to match each word form.

7. two and seven tenths

8. thirty-four and five hundredths

9. twenty-eight hundredths

Write each decimal in word form.

10. 18.4

11. 9.15

12. 4.004

Lesson 1.2 ~ Rounding Decimals

Round each decimal to the place value of the digit that is underlined.

13. 52.9$\underline{8}$7

14. 3.$\underline{8}$43

15. 6.35$\underline{8}$6

16. 13.0$\underline{9}$8

17. 8.$\underline{9}$71

18. $\underline{6}$.7

Round each decimal to the place value of the digit that is underlined.

19. 93.0<u>0</u>9

20. 4.<u>0</u>07

21. 5<u>9</u>.99

22. Nadia wants to buy a dress that costs $28.75. Approximately how much money should she bring to the nearest dollar?

23. Elliot buys candy rings for $1.38. How much, to the nearest dime, should he give the cashier?

Lesson 1.3 ~ Measuring in Centimeters

24. The height of a salt shaker is 14 centimeters and 5 millimeters. Write this measurement as a decimal with centimeter units.

Measure the length of each line to the nearest tenth of a centimeter.

25. ─────────────────

26. ────────────────────

27. ────

Measure the length of each line to the nearest half centimeter.

28. ──────────────

29. ─────

30. ──────────────────

31. Use the picture of the lizard.
   a. Estimate the length of the lizard to the nearest tenth of a centimeter.
   b. Measure the length of the lizard to the nearest tenth of a centimeter.
   c. How far off was your estimate, to the nearest tenth of a centimeter?

Lesson 1.4 ~ Ordering and Comparing Decimals

Replace each ⬤ with <, > or = to make a true sentence.

32. 4.3 ⬤ 4.40

33. 70.2 ⬤ 70.23

34. 44.9 ⬤ 44.09

35. 9.53 ⬤ 9.530

36. 8.01 ⬤ 8

37. 11.101 ⬤ 11.0101

Put each set of numbers in order from least to greatest.

38. 9.09, 9.9, 9.009, 9

39. 0.88, 0.8, 0.842, 0.884

40. On Monday it snowed 2.125 inches in Bend. On Tuesday it snowed 2.1 inches and on Wednesday it snowed an additional 2.13 inches.
    a. On what day did it snow the most?
    b. On what day did it snow the least?

Lesson 1.5 ~ Estimating with Decimals

Estimate each sum, difference or product using rounding to the nearest whole number.

41. 9.6 + 3.4

42. 11.74 – 10.442

43. 6.62 × 4.3

44. 8.98 – 2.76

45. 23.45 + 6.25

46. 10.1 × 10.92

47. Joshua bought three bags of dog food. One bag weighed 4.89 pounds. The other two bags weighed 5.4 pounds and 2.51 pounds. What was the approximate weight of all three bags combined?

Estimate each quotient using compatible numbers.

48. 14.7 ÷ 7.04

49. 43.9 ÷ 5.32

50. 92.13 ÷ 9.45

51. Carlita used 8.14 gallons of gasoline to drive 234.6 miles. Estimate the number of miles she can drive with one gallon of gasoline.

Lesson 1.6 ~ Adding and Subtracting Decimals

Find each sum or difference.

52. 9.1 + 2.4

53. 1.981 – 0.682

54. 31.321 – 28.198

55. 6.608 + 9.44

56. 6.71 + 3.32

57. 4.564 + 8.5

58. 7.902 – 3.42

59. 4.86 – 2.9

60. 14 + 2.57

61. Brooklyn had a water bottle that contained 16.9 ounces of water. She drank 8.75 ounces of the water. How many ounces were left?

62. Nevaeh measured 2.5 cups of flour and 1.75 cups of sugar into a bowl. How many cups of ingredients did she have altogether in the bowl?

TIC-TAC-TOE ~ CHECKBOOK REGISTRY

You received $300 to spend on gifts for family and friends. Create a checkbook registry like the one below. Write $300.00 as your first deposit entry.

Use local advertisements to cut out items you would like to buy. Write each item and its price into the registry. Subtract the price from the running total.

Example:

Original Deposit		+$300.00
Shoes	$49.96	−$49.96
		$250.04
Portable DVD Player	$89.99	−$89.99
		$160.05
Curling Iron	$10.99	−$10.99
		$149.06
MP3 Player	$79.99	−$79.99
		$69.07
Digital Frame	$68.99	−$68.99
		$0.08

RULES: You must "buy" at least five gifts. (You do not really have to buy them.) You must purchase gifts until you have less than one dollar left.

Create a poster with copies of the advertised prices of the gifts you "purchased" and a copy of your checkbook registry.

TIC-TAC-TOE ~ PURCHASE SPREADSHEET

Step 1: Copy and complete the table below. For example, if your favorite movie is "What's Up," write that in the "My Favorite" column beside "DVD." Find the cost of each item at a store, on the internet or in a newspaper advertisement. Record the price in the "Price" column.

Step 2: Open a spreadsheet program on a computer. Create column headings for a chart as listed below.

In cell A1 type: Category
In cell B1 type: My Favorite
In cell C1 type: Price
In cell D1 type: Price for Three

Category	My Favorite	Price	Price for Three
Book			
DVD			
Board game			
Snack			
CD			
Video game			
Beverage			
Food			

Continued on next page

Step 3: Type the categories from **Step 1** in cells A2 through A9. You can adjust the width of the cell for category columns by placing your cursor on the line between boxes A and B at the top of the spreadsheet. As the cursor touches the line, it becomes a "T" shape with arrows pointing in either direction. Click and drag the line to the right to make the cells for column A wider.

Step 4: Type the name of your favorite item in each category from **Step 1** in cells B2 through B9. Make the column wider following the procedure in **Step 3**.

Step 5: Enter the price for each item in the B column using a decimal point in cells C2 through C9.

> *Example:* Type "23.95" for a price that reads $23.95.

> *Hint:* To create cells displaying dollar amounts, highlight the cells which need a dollar sign, go to "Format" on the menu bar and select "cells." On the left hand list select "numbers" and choose "currency" from the list under the word "category."

Step 6: Type "= 3*C2" in cell D2 and press the "enter" key. This tells the spreadsheet program to multiply the value in C2 by 3. As you type "C2" a blue box will highlight C2. Check the highlighted box to make sure you have selected the price you want to multiply by 3. After pressing "enter," a price should appear in D2 that is three times the amount of the price in C2. *Note:* In a spreadsheet, the * symbol represents multiplication.

Step 7: Continue the process in **Step 6** with the cells in column D. Use the correct cell number.

Step 8: When you have prices for three of each item, click on cell C10. Find the button with the Σ sign at the top of the spreadsheet. Click on the Σ and "=SUM(C2:C9)" will appear in cell C10. Press enter. This gives the total for all amounts in column C. Click on cell D10 then the Σ sign. The function "=SUM(D2:D9)" will appear in D10. Press enter. This gives the total for all amounts in column D.

Step 9: Highlight all cells with typing in them. Find the button at the top of the spreadsheet to create borders. Border all lines that are in the area you have highlighted. Print the spreadsheet by choosing "print selection" while the cells remain highlighted.

Step 10: Answer the following questions in complete sentences on a separate sheet of paper. Attach the answers to the spreadsheet.

1. Which single item costs the most?
2. If you buy three of your favorite books, one for yourself and two friends, how much would you spend?
3. An advertisement has a coupon for "Buy three CDs, Get $4.99 Off." What would three CDs cost if you used the coupon?
4. List the three cheapest items. What is the total cost if you buy three of each of these?
5. List the four most expensive items. What is the total cost if you buy three of each of these?
6. Many businesses and individuals budget their money with a spreadsheet. Why do you think they use spreadsheets?

CAREER FOCUS

LUIS
ESCROW OFFICER

I am an escrow officer. I deal with legal documents and work with the county to sell and buy homes. In one home purchase transaction I work with realtors, lenders, mortgage brokers, buyers and sellers. My company is a neutral third party that collects information and funds. We keep them until it is time to complete the transaction. Escrow officers are important members of any real estate transaction.

I use basic math in my profession for many things. I add and subtract money that comes in or goes out. I also use ratios to prorate items in transactions. Prorating means to find a part of the whole amount. I prorate taxes and, sometimes, rent for my clients. An example of this is calculating how much tax the buyer and the seller would each need to pay when a house is purchased part way through the year. The amount each one owes is determined with proportions. Math is important in my job to make sure that people are not paying too little or too much.

You must have a high school diploma or equivalent to become an escrow officer. People usually work in an escrow office for 3 or 4 years to gain valuable experience before they become officers.

An escrow officer's salary ranges from $38,000 to $58,000 per year. The salary depends on how experienced they are and how many clients they have.

One thing I like about my profession is the variety of transactions. No sale or refinance is the same. This keeps me busy each day and makes my job interesting. One day I might have a smooth transaction without any problems. Another day I may have a transaction with a lot of obstacles. When there are problems I am like a detective. I try to find out exactly what the problems are and find solutions for them. Another thing that I like about my job is helping people achieve the American Dream of owning a home. This is very rewarding to me.

CORE FOCUS ON DECIMALS & FRACTIONS

BLOCK 2 ~ MULTIPLYING AND DIVIDING DECIMALS

LESSON 2.1   MULTIPLYING BY 2-DIGIT NUMBERS --- 34

LESSON 2.2   MULTIPLYING DECIMALS --- 38

                EXPLORE! SMART SHOPPER

LESSON 2.3   DIVIDING BY 1-DIGIT NUMBERS -- 43

                EXPLORE! BEADED NECKLACES

LESSON 2.4   DIVIDING BY 2-DIGIT NUMBERS -- 49

                EXPLORE! MAGAZINE SUBSCRIPTIONS

LESSON 2.5   DIVIDING DECIMALS BY WHOLE NUMBERS ----------------------------------- 54

LESSON 2.6   DIVIDING DECIMALS BY DECIMALS --- 58

REVIEW   BLOCK 2 ~ MULTIPLYING AND DIVIDING DECIMALS ------------------------- 63

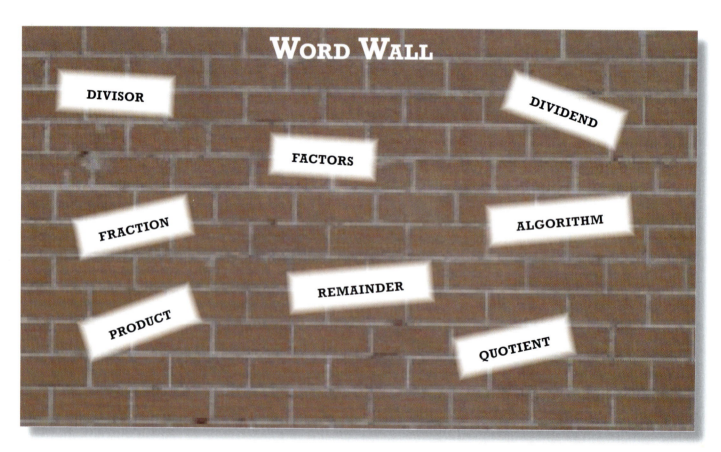

BLOCK 2 ~ MULTIPLYING AND DIVIDING DECIMALS
TIC-TAC-TOE

QUANTITY PRICING

Take prices of items in large quantities and compare them to prices of the same items bought in smaller quantities.

See page 48 for details.

MULTIPLYING METHODS

Explore two different ways to multiply multi-digit numbers.

See page 42 for details.

COLOR BY NUMBERS

Create a picture and divide into pieces with multiplication and division of decimals. Answers equal colors.

See page 65 for details.

DECIMAL POETRY

Write a poem about multiplying decimals. Write another poem about dividing by decimals.

See page 62 for details..

A PENNY SAVED

Decide between two different ways of saving money.

See page 41 for details.

AIRLINE MILES

Design an airline mile plan.

See page 37 for details.

RACING THE CLOCK

Time students racing short distances versus long distances.

See page 53 for details.

DRAWING DIMENSIONS

Measure the dimensions of two different rooms in your house using centimeters. Draw each to scale.

See page 53 for details.

TEACHER, TEACHER

Create a "Dividing Decimals" manual that teaches other students how to divide decimals.

See page 62 for details.

MULTIPLYING BY 2-DIGIT NUMBERS

Find products of expressions involving multi-digit whole numbers.

Johnny has four jars that each hold twenty-eight marbles. He wants to find the total number of marbles in the jars. He could use repeated addition (28 + 28 + 28 + 28) or multiplication (28 × 4). Multiplication is a process to solve problems involving repeated addition. Multiplication problems can be written horizontally or vertically.

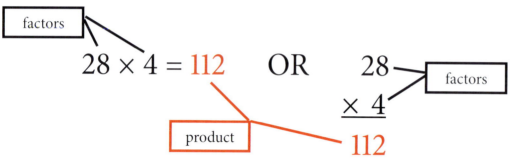

factors

$$28 \times 4 = 112 \qquad OR \qquad \begin{array}{r} 28 \\ \times\ 4 \\ \hline 112 \end{array}$$

product

factors

Factors are two or more numbers that can be multiplied together to find a product. The **product** is the answer to a multiplication problem. Pay careful attention to the place value of numbers as you are multiplying.

EXAMPLE 1

What is the product of 18 × 5?

SOLUTION

Write the problem vertically with the largest factor on top. Line up by place value.

Tens	Ones
1	8
×	5

Multiply the ones: 5 × 8 = 40. Regroup (40 = 4 tens + 0 ones) and carry the 4 tens to the tens column. Record the 0 ones.

Tens	Ones
4	
1	8
×	5
	0

The 4 is from 40.

The 0 is from 40.

Multiply by the tens: 5 × 10 = 50

Add the tens that were carried.
5 tens (50) + 4 tens (40) = 90.
Record the 9 tens.

18 × 5 = 90

Tens	Ones
4	
1	8
×	5
9	0

The 9 is from 90.

Example 1 shows the standard algorithm for multiplication. An algorithm gives steps for calculating values. The standard algorithm for multiplication gives steps for finding products.

STANDARD ALGORITHM FOR MULTIPLICATION

1. Multiply right to left. Start with the ones column and move to tens column (and so on).
2. Regroup when possible.
3. Add any partial products to find the final product.

When multiplying with larger factors, using the standard algorithm can help you keep your answer organized.

EXAMPLE 2

Tatiana loaded boxes of books into a warehouse. She loaded 38 boxes per hour for 24 hours this week. How many boxes of books did she load into the warehouse this week?

SOLUTION

Write the problem vertically with the largest number on top. Line up by place value.

Hundreds	Tens	Ones
	3	8
×	2	4

Multiply the ones (4 from 24 by the 8 from 38).
$4 \times 8 = 32$

Regroup (32 = 3 tens and 2 ones) and carry the 3 tens to the tens column. Record the 2 ones.

Hundreds	Tens	Ones
	3	
	3	8
×	2	4
		2

Multiply the ones from the second factor by the tens from the first factor: (4 from 24 by the 3 from 38) $4 \times 3 = 30$ (3 tens) = 120 (12 tens).

Add the tens that were carried.
3 tens (30) + 12 tens (120)= 150.
Record the 5 tens. Carry 1 hundred and bring down.

Hundreds	Tens	Ones
1	3	
	3	8
×	2	4
1	5	2

Multiply the tens from the second factor by the ones from the first factor (2 from 24 by the 8 from 38) 2 tens (20) $\times$ 8 = 160 (1 hundred and 6 tens). Record the 6 tens. Carry the 1 hundred.

Hundreds	Tens	Ones
1		
	3	8
×	2	4
1	5	2
	6	0

There are no ones. Record a 0 in the ones answer column.

Continued on next page.

EXAMPLE 2
SOLUTION
(CONTINUED)

Multiply the tens from the second factor by the tens from the first factor: (2 from 24 by the 3 from 38).
2 tens (20) × 3 tens (30) = 600 (6 hundreds)

Add the hundred that was carried.

Hundreds	Tens	Ones
1		
	3	8
×	2	4
1	5	2
+ 7	6	0

6 hundreds
+ 1 hundred
7 hundreds

Add the two partial products together.

152
+ 760
912

Tatiana loaded 912 boxes of books into the warehouse this week.

EXERCISES

Find each product.

1. 17
    × 6

2. 29
    × 4

3. 34
    × 7

4. 51 × 3

5. 43 × 5

6. 67 × 8

7. Stefán wrote songs for 45 minutes each day for 8 days. How many minutes in all did he write songs?

8. Lee bikes 24 miles each day. After 7 days, how many total miles has he biked?

9. 32
    × 14

10. 72
    × 43

11. 81
    × 26

12. 59 × 13

13. 65 × 39

14. 84 × 46

15. Bella uses 56 beads to create a necklace. She made 23 necklaces for a craft show. How many beads did she use in all?

16. Rafael had 18 boards that measured 76 centimeters each. How many centimeters of board did he have altogether?

17. Anna canned 12-ounce jars of applesauce. If she canned 16 jars, how many ounces of applesauce does she have?

18. Jamal shoots 85 free throws each day. In the month of December how many free throws will he shoot?

19. 186
　　× 23

20. 276
　　× 54

21. 321
　　× 39

22. Brittany made $485 each month. In 12 months, how much would she make?

23. Darien swam 52 meters each day for 6 days. Then he swam 68 meters each day for 8 days. How many total meters did he race in all?

24. Imelda measured and cut 24 pieces of ribbon. Each piece of ribbon was 84 centimeters long. When she made the ribbons into bows, she realized she had made each piece of ribbon 11 centimeters too long. After cutting off the extra ribbon on each bow, how much ribbon did she use for 24 bows?

REVIEW

Put each group of measurements in order from least to greatest.

25. 8.3 *cm*, 7.8 *cm*, 7.2 *cm*, 8.6 *cm*

26. 5.25 *m*, 5.025 *m*, 5.2 *m*, 5.205 *m*

27. 0.3 *cm*, 3 *cm*, 3.3 *cm*, 0.33 *cm*

28. 9.78 *km*, 9.7 *km*, 9.708 *km*, 9.8 *km*

29. Bonnie walked 3.2 kilometers from her house to her friend's house. Along the way she stopped at the library. The library was 1.4 kilometers before her friend's house. How far did Bonnie walk to get to the library?

30. Chad collected 86.5 ounces of baby food for a food drive on Monday. On Tuesday, he collected 62.3 ounces of baby food. How many total ounces of baby food did he collect?

TIC-TAC-TOE ~ AIRLINE MILES

Many airlines provide airline miles that can be redeemed for a variety of items. You can earn miles in many ways. For example, you may get bonus miles for signing up. Some airlines give bonus miles once per year. You may also get bonus miles for flying – sometimes for the cost of the ticket and other times for the number of miles you fly.

Research at least three different plans. Design an airline miles plan that would be competitive with the companies you researched. What is the cost? How do people earn miles? Provide incentives. What can people earn with their miles? How many miles does it take to earn rewards?

Present your airline plan to an adult. If the adult has any suggestions for changes to make it more appealing to someone searching for an airline plan, adapt your plan as needed.

MULTIPLYING DECIMALS

Find products of expressions involving decimals.

Yana had make-your-own banana splits for her birthday party. She had toppings at home, but she purchased three half-gallons of ice cream and 3.6 pounds of bananas. How much did these items cost?

Ice cream: $3.55 per half gallon	Bananas: $0.40 per pound

MULTIPLYING WITH DECIMALS

1. Multiply as if the numbers were whole numbers.
2. Count the number of places after the decimal points in each factor.
3. Count the same number of places in the product starting from the right and moving left.
4. Insert zeros, if needed, to hold place value. Delete zeros that are not necessary.
5. Put the decimal point where you stop counting in the answer.

EXAMPLE 1

How much did Yana spend on three half-gallons of ice cream if each half-gallon cost $3.55?

SOLUTION

Write the problem. $3 \times \$3.55$

Use decimal models to visualize.

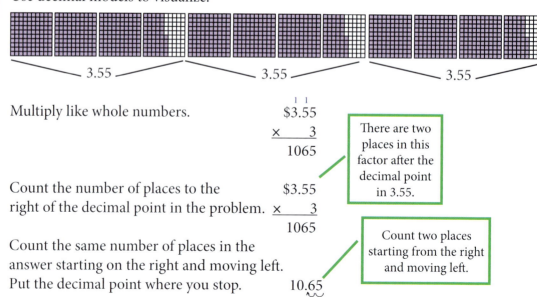

3.55   3.55   3.55

Multiply like whole numbers.

$$\begin{array}{r} \overset{1\ 1}{\$3.55} \\ \times \quad 3 \\ \hline 1065 \end{array}$$

There are two places in this factor after the decimal point in 3.55.

Count the number of places to the right of the decimal point in the problem.

$$\begin{array}{r} \$3.55 \\ \times \quad 3 \\ \hline 1065 \end{array}$$

Count the same number of places in the answer starting on the right and moving left. Put the decimal point where you stop.

10.65

Count two places starting from the right and moving left.

Yana spent $10.65 on ice cream.

EXAMPLE 2	**How much did Yana spend on 3.6 pounds of bananas if they cost $0.40 per pound?**
SOLUTION	

Write the problem.   3.6 × $0.40

Use decimal models to visualize.

3.6

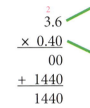

> 0.40

Color decimal models vertically to match the first decimal.

Color the same decimal models horizontally with the second decimal.

Where the two colors overlap is the answer.

0.40 + 0.40 + 0.40 + 0.24 = 1.44

Multiply like whole numbers. Count the number of digits to the right of the decimal points in both numbers in the problem.

$$\begin{array}{r} \overset{2}{3.6} \\ \times\ 0.40 \\ \hline 00 \\ +\ 1440 \\ \hline 1440 \end{array}$$

There is one place after the decimal point in the factor 3.6.

There are two places after the decimal point in the factor 0.40.

Count the same number of places in the answer starting on the right and moving left. Put the decimal point where you stop.   1.440

Count three places starting from the right and moving left.

Drop any zeros at the end of the decimal.   1.440 = 1.44

Yana spent $1.44 on bananas.

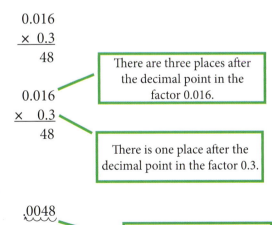

EXAMPLE 3	**Find the value of 0.3 × 0.016.**
SOLUTION	

Multiply like whole numbers.

$$\begin{array}{r} 0.016 \\ \times\ 0.3 \\ \hline 48 \end{array}$$

Count the number of digits to the right of the decimal points in both numbers in the problem.

$$\begin{array}{r} 0.016 \\ \times\ 0.3 \\ \hline 48 \end{array}$$

There are three places after the decimal point in the factor 0.016.

There is one place after the decimal point in the factor 0.3.

Count the same number of places in the answer starting on the right and moving left. Insert zeros to hold the missing places. Put the decimal point where you stop.   .0048

Count four places starting from the right and moving left.

0.3 × 0.016 = 0.0048

Jay gets advertisements for the local grocery stores in his newspaper. He wants to save the most money possible. He figures out which stores he should buy which items from. Here is part of his list and the stores with sales on those items.

ABC Grocery
Grapes: $1.39 per pound
Cherry tomatoes: $0.98 per pound
Fuji apples: $1.29 per pound
Deli-sliced ham: $3.99 per pound
Chicken breasts: $1.69 per pound
Whole chicken fryer: $0.69 per pound
Beef round roast: $1.98 per pound
Chocolate Blitz bars: $0.59 each
Fizzy Pop: $0.70 each

ZYX Grocery
Grapes: 3 pound container for $4.99
Cherry tomatoes: 3.5 pound container $3.96
Fuji apples: 3.25 pound bag for $4.25
Deli-sliced ham: 1.75 pound package for $5.50
Chicken breasts: 4 pound bag for $5.99
Whole chicken fryer: $5.50 for 5 pounds
Beef round roast: 4.5 pound roast for $9.23
Chocolate Blitz bars: $6.40 for bag of 12
Fizzy Pop: $9.99 for 15 pack

From which store should Jay buy grapes?

Step 1: Multiply the price at ABC Grocery ($1.39) by the number of pounds he would buy if he bought grapes at ZYX Grocery (3 pounds).

Step 2: Round to the nearest penny or hundredth if the total price when multiplied ends with more than two places after the decimal point. This is the price for the same amount of grapes at ABC Grocery.

Step 3: Compare prices. Which store has the lower price?

Step 4: Use the procedure from **Steps 1-3** for each item. Decide which store has the best price for each item on Jay's list.

EXERCISES

Find each product.

1. 4.1 × 3

2. 5 × 3.33

3. 7 × $2.17

4. 6.312 × 4

5. $4.89 × 5

6. 6 × 9.765

7. Celia's cell phone company charges $0.39 per minute if she goes over her allotted minutes for the month. Last month she went over by 15 minutes. How much extra did she owe on her bill?

8. Kenyan bought three pairs of pants that were on sale for $12.99 per pair. How much did he pay altogether?

Find each product.

9. 3.2×7.4

10. 2.5×6.6

11. 12.3×2.8

12. 10.45×4.1

13. 5.72×3.4

14. 11.5×7.62

15. 0.5×0.9

16. 0.3×0.46

17. $\$0.35 \times 0.8$

18. Hakeem bought 4.2 pounds of almonds. They cost $2.45 per pound. How much did Hakeem pay for the almonds?

19. The price tag under the 11.5 ounce bag of chips at the grocery store says $0.24 per ounce. How much does the bag of chips cost?

20. 0.012×3

21. 2.1×0.025

22. 4.12×0.0065

REVIEW

List each set of numbers from least to greatest.

23. 0.1, 0.09, 0.05

24. 0.6, 0.62, 0.58

25. 1.31, 1.089, 1.4

Estimate the value of each expression using rounding to the nearest whole number.

26. $4.3 + 9.9$

27. $22.553 - 10.39$

28. 3.6×6.54

29. $15.442 - 8.92$

30. 7.2×2.14

31. $7.57 + 8.62$

TIC-TAC-TOE ~ A PENNY SAVED

Decide which choice for earning money is the best if you earn money for 20 days, 30 days or 40 days. Support your answer with words and/or symbols.

CHOICE #1: You start with $0.01. You make double what you had on the previous day. Add that to the amount you had the previous day.

　　Example: Day 1 = $0.01

　　　　　　Day 2 = ($0.01 × 2 = $0.02) + Day 1 total ($0.01) = $0.03

　　　　　　Day 3 = ($0.03 × 2 = $0.06) + Day 2 total ($0.03) = $0.09

CHOICE #2: You start with $0.01. On Day 10, you make 100 times that amount. Add this to the amount you had on Day 1. On Day 20, you make 100 times the amount you had on Day 10. Add this to the amount you had on Day 10. On Day 30, you make 100 times the amount you had on Day 20. Add this to the amount you had on Day 20.

　　Example: Day 1 = $0.01

　　　　　　Day 10 = ($0.01 × 100 = $1.00) + Day 1 amount ($0.01) = $1.01

TIC-TAC-TOE ~ MULTIPLYING METHODS

Write 10 multi-digit multiplication problems. Use the Grid Method for five of the problems. Use the Lattice Method five of the problems. Check your work by using the standard algorithm (taught in **Lesson 2.1**). Write a paragraph explaining the method you like the best and why it is your favorite. Write a paragraph to explain how this method would work when multiplying decimals.

Grid Method: 45 × 13

Step 1: Split each factor by place value. (45 = 40 + 5; 13 = 10 + 3)

Step 2: Write each part of the first factor across the top of the grid (40 & 5). Write each part of the second factor down the grid (10 & 3).

Step 3: Multiply each part (40 × 10 = 400 (answer goes in that grid box) and so on).

	40	5
10	400	50
3	120	15

Step 4: Add all of the partial products together (400 + 50 + 120 + 15 = 585).

Step 5: Check your work using the standard algorithm.

Lattice Method: 45 × 13

Step 1: Create a grid for the numbers (2 boxes by 2 boxes because this problem has two 2-digit factors.

Step 2: Split each box in half with a diagonal line (see below).

Step 3: Multiply each number (4 × 1 = 4; 5 × 1 = 5, 4 × 3 = 12, 5 × 3 = 15). The tens of an answer go in the top portion of the box, the ones of an answer go in the bottom portion of a box.

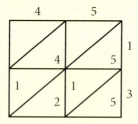

Step 4: Add answers in lattice diagonally starting from the right.

   Ones: 5 + 0 = 5   Tens: 5 + 1 + 2 = 8   Hundreds: 4 + 1 = 5

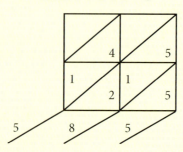

Step 5: Check your work using the standard algorithm.

DIVIDING BY 1-DIGIT NUMBERS

 Find quotients of expressions where whole numbers are divided by 1-digit whole numbers, including remainders.

Division is the process of splitting something into equal parts. Division problems can be written two ways:

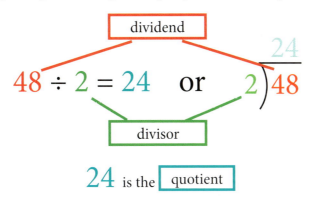

24 is the quotient

The **dividend** is the number you are dividing. The **divisor** is the number you are dividing by. The **quotient** is the answer.

Dividing means you look at how many groups of one number fit into another number.

$48 \div 2$ can be read:
"How many groups of 2 fit into the number 48?"
OR
"How many would be in each group if you divided 48 equally between 2 groups?"

You can use the relationship between multiplication and division to make sure your answers are correct.

$$2 \times 24 = 48 \qquad\qquad 48 \div 24 = 2$$

$$24 \times 2 = 48 \qquad\qquad 48 \div 2 = 24$$

These four facts are called a fact family.

Taylor is making necklaces for 4 people. She wants each necklace to have the same number of beads of each color. Listed below are the beads Taylor has.

52 red beads   76 white beads   48 brown beads
68 black beads   64 clear beads   80 blue beads

Step 1: Set out 52 Base-Ten Blocks to model the 52 red beads.

5 tens sticks = 50   2 ones cubes = 2   50 + 2 = 52

Step 2: Separate the 5 tens sticks into 4 piles to show the 4 necklaces Taylor is making.

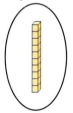 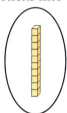 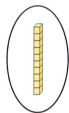 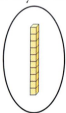

Extra stick

Step 3: Trade the extra tens stick for 10 ones cubes. Add these to the 2 ones cubes you had to start with.

1 ten stick = 10 ones   10 ones + 2 ones = 12 ones

Step 4: Separate the 12 ones cubes into the 4 piles to show the 4 necklaces Taylor is making.

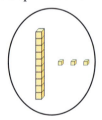 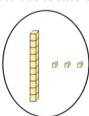 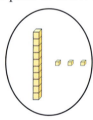 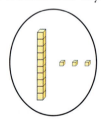

Each necklace will have 13 red beads.

Step 5: Use **Steps 1-5** above and Base-Ten Blocks to figure out how many…
    a. white beads Taylor will have for each necklace.
    b. brown beads Taylor will have for each necklace.
    c. black beads Taylor will have for each necklace.
    d. clear beads Taylor will have for each necklace.
    e. blue beads Taylor will have for each necklace.

In the Explore!, you modeled division with Base-Ten blocks. Division can be shown without manipulatives using five steps:

DIVIDE – MULTIPLY – SUBTRACT – DROP DOWN – REPEAT

EXAMPLE 1

What is the quotient of 58 ÷ 2?

SOLUTION

<u>Divide</u>. Begin division with the digit in the largest place value in the dividend (<u>5</u>8). 5 ÷ 2

$$2\overline{)58}$$

<u>Multiply</u>. How many times can the divisor go into the number without going over? 2 × _____ = _____ (a number close to 5) 2 × 2 = 4. Write a 2 in the tens place of the quotient. Write the number 4 below the 5.

$$2\overline{)\underset{4}{58}}\quad ^{2}$$

<u>Subtract</u>. Subtract 5 – 4 = 1. Write 1 below the number 4 in the tens column. Make sure your partial difference (1) is less than your divisor (2). If it is not, a mistake has been made.

$$2\overline{)58}\quad ^{2}$$
$$\underline{-4}$$
$$1$$

<u>Drop Down</u>: Bring the next number in the dividend (8) down with the partial difference and keep the 8 in the ones column.

$$2\overline{)58}\quad ^{2}$$
$$\underline{-4}\ \downarrow$$
$$1\ \ 8$$

<u>Repeat</u>.
1. Divide: 18 ÷ 2.

2. Multiply: 2 × _____ = _____ (18 or a number close to 18) 2 × 9 = 18.

3. Subtract: 18 – 18 = 0.

4. Drop Down: When the last partial difference is 0, the divisor divides evenly into the dividend.

58 ÷ 2 = 29

$$2\overline{)58}\quad ^{2\ 9}$$
$$\underline{-4}\ \downarrow$$
$$1\ \ 8$$
$$\underline{-1\ \ 8}$$
$$0$$

CHECK YOUR ANSWER. Use the relationship between multiplication and division.
58 ÷ 2 = 29 so
29 × 2 = 58

A **remainder** is the number that is left over when the division problem is completed. It is always written after the whole number in the quotient. It can be written with an R for remainder or as a fraction. A **fraction** is a number that represents part of a whole number. It is written $\frac{numerator}{denominator}$.

For example: 58 ÷ 7 = 8 R2   or $58 \div 7 = 8\frac{2}{7}$

The remainder is written $\frac{remainder}{divisor}$.

EXAMPLE 2

Terri made 395 chocolate cake pops for her catering business. She can put them into groups of 8 in jars to decorate tables at a party. How many jars will she need?

SOLUTION

<u>Divide</u>: Begin division with the digit in the largest place value in the dividend (<u>3</u>95). Can 8 go into 3? No. Place an X in the box above 3 in the hundreds column of the quotient. How many times does 8 divide into 39 without going over?

<u>Multiply</u>: $8 \times \underline{4} = 32$ (a number close to 39). Write the factor (4) in the quotient above the 9 in the tens column. Write the product (32) on the line below 39.

<u>Subtract</u>: Subtract 39 − 32 = 7. Write the difference (7) below the number 32. Make sure your partial difference (7) is less than your divisor (8). If it is not, a mistake has been made.

<u>Drop Down</u>: Bring the next number (5) in the dividend down to the right of the partial difference in the ones column.

It makes sense to write this remainder as R3 because there are 3 cake pops left after Terri is finished arranging them in jars.

<u>Repeat</u>:
1. Divide 75 by 8.

2. Multiply $8 \times \underline{9} = 72$. The factor (9) goes in the quotient above the 5 (ones column).

3. Subtract the answer from 75. (75 − 72 = 3)

4. Drop Down: Nothing to drop down from dividend.

5. The remainder is written after the whole number part of the quotient as R3.

Terri needs 49 jars for her cake pops. She will have 3 cake pops left over.

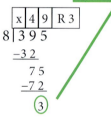

Always check your answer by using multiplication.

If 395 ÷ 8 = 49 R3, then $49 \times 8 + 3 = 395$.

In **Example 2** the remainder was written as a whole number (R3). Sometimes it makes more sense to write the remainder as a fraction, especially when working with measurements. For example, 325 inches ÷ 4 = 81 R1 or $81\frac{1}{4}$ inches. The measurement $81\frac{1}{4}$ inches makes more sense in this situation than 81 R1.

46   *Lesson 2.3 ~ Dividing by 1-Digit Numbers*

EXERCISES

Find each quotient. Write any remainder as R__.

1. 46 ÷ 3

2. 4)̅76̅

3. 65 ÷ 4

4. 5)̅84̅

5. 3)̅77̅

6. 96 ÷ 8

Find each quotient. Label your answer.

7. Rachel set up a banquet hall for 56 people. She put 4 people at each table. How many tables did she need for the banquet?

8. Jovanna had 74 flowers. She made bouquets of equal amounts for 6 people. How many flowers did each person receive? How many flowers were left over?

9. There were 174 people at a presentation. They were seated in 9 equal rows in the room. How many people sat in each row? How many people were left?

Find each quotient. Write any remainder as a fraction.

10. 3)̅426̅

11. 205 ÷ 6

12. 5)̅324̅

13. 351 ÷ 2

14. 143 ÷ 4

15. 7)̅572̅

Find each quotient. Label your answer.

16. Carl measured 75 yards from start to finish for a relay race. He split the total yards into 4 equal lengths. How many yards long was each portion of the race?

17. The Fruit Basket Company had 759 pounds of fruit for 7 different organizations' fruit baskets. Every organization was given the same amount of fruit. How many pounds of fruit did each organization receive?

18. Macy had 126 minutes of free time. She wanted to complete 4 activities and spend an equal amount of time on each activity. How many minutes did Macy have for each activity?

Find each quotient. Write any remainder as R__ and as a fraction.

19. 8)̅292̅

20. 90 ÷ 7

21. 982 ÷ 5

22. 81 ÷ 3

23. 4)̅67̅

24. 8)̅270̅

25. 6)̅802̅

26. 3)̅83̅

27. 434 ÷ 7

Find each product.

28. 7.5 × 3

29. 4.1 × 3.4

30. 0.6 × 2.3

31. 9.2 × 1.82

32. 6 × $7.75

33. 0.013 × 3.5

Find each sum.

34. 2.94 + 3.22

35. 9.32 + 8.091

36. 44.229 + 21.981

TIC-TAC-TOE ~ QUANTITY PRICING

Step 1: Choose an online grocery store (or store if you choose to visit one). Price the following items in two different quantities (i.e., an 18 pack of toilet paper and a 4 pack of toilet paper).

ITEM	QUANTITY #1 & PRICE	QUANTITY #2 & PRICE
Toilet paper		
Paper Towels		
Flour (lbs.)		
Sugar (lbs.)		
Napkins		
Soda Pop		
Diapers		

Step 2: Divide the price of each item by the quantity. Toilet Paper: 18.99 ÷ 36 (rolls) = $0.5275
3.99 ÷ 4 (rolls) = $0.9975

Step 3: Round each decimal to nearest penny (hundredth).
Toilet Paper: 18.99 ÷ 36 (rolls) = $0.5275 = $0.53 per roll
3.99 ÷ 4 (rolls) = $0.9975 = $1.00 per roll

Step 4: Decide which quantity of each item would be the best price for purchasing.

Step 5: Are there items in the table above that are best to buy in larger quantities? Are there household situations that require larger quantities or smaller quantities of an item? Are there times that buying in smaller quantities would be more beneficial for you or your family? Write a letter to the person who shops for groceries at your house explaining your thinking.

Find quotients of expressions where whole numbers are divided by multi-digit whole numbers, including remainders.

EXPLORE!   **MAGAZINE SUBSCRIPTIONS**

For her club fundraiser, Stacy has to sell magazines. Each magazine subscription costs a different amount. Stacy has a goal of selling at least 80 subscriptions for her club.

Famous - $13 per year subscription   *Working with Wood* - $11 per year subscription

Sports For All - $20 per year subscription   *Scrapbooking Ideas* - $15 per year subscription

Stacy made $273 by selling subscriptions to *Famous*. How many subscriptions did she sell of this magazine?

Sean, Mia, Jabar and Angelina chose four different ways to solve this problem:

Estimate and Guess	Subtraction	Addition	Standard Algorithm

Sean chose a number to multiply by 13. $13 \times 5 = 65$ He could subtract that from 273. $273 - 65 = 208$
 	
He was still left with a large number so he tried a larger estimate $13 \times 10 = 130$ and subtracted this. $208 - 130 = 78$
 	
He saw that 78 is not much more than 65 so he tried $13 \times 6 = 78$ and subtracted this. $78 - 78 = 0$
 	
He added all the factors he had multiplied by 13: $5 + 10 + 6 = 21$
	Mia chose to subtract 13 until she got to 0.
	
	273 130
	−13 −13
	260 117
	−13 −13
	247 104
	−13 −13
	234 91
	−13 −13
	221 78
	−13 −13
	208 65
	−13 −13
	195 52
	−13 −13
	182 39
	−13 −13
	169 26
	−13 −13
	156 13
	−13 −13
	143 0
	−13
	130
	
	She counted how many 13s she subtracted.
		Jabar chose to add groups of ten "13s". If ten more could not fit, he added individual 13s until he got to 273.
		
		He kept track of how many 13s by writing the number of 13s above the addend. He added up how may 13s he had used to find the answer.
		
		$10 \times 13 = 130$
		
		10 10 1 21
		$130 + 130 + 13 = 273$
			Angelina chose to use the standard algorithm.
			
			     21
			13)273
			  −26
			    13
			   −13
			     0

Stacy sold 21 subscriptions to *Famous*.	Stacy sold 21 subscriptions to *Famous*.	Stacy sold 21 subscriptions to *Famous*.	Stacy sold 21 subscriptions to *Famous*.

Step 1: Which method do you like best? Why?

Step 2: Here are the total amounts Stacy made for each magazine title. Choose a method shown on the previous page. Use the method to figure out how many of each subscription Stacy sold.
    a. *Sports for All*: $520
    b. *Working with Wood*: $154
    c. *Scrapbooking Ideas*: $285

Step 3: Did Stacy sell enough subscriptions to meet her goal? Support your answer with words and/or symbols.

Dividing by 2-digit divisors is much like dividing by 1-digit divisors. However, instead of dividing into just the first digit of the dividend, you will start by dividing into the first two digits (or more) of the dividend.

DIVIDE – MULTIPLY – SUBTRACT – DROP DOWN – REPEAT

EXAMPLE 1

What is the quotient of 168 ÷ 12 ?

SOLUTION

Divide: 1 cannot be divided by 12. Put an X over the 1 in the quotient. 16 can be divided by 12.

$$\begin{array}{r} \text{x} \\ 12\overline{)168} \end{array}$$

Multiply: 12 × <u>1</u> = 12 (closest number to 16). Write the factor (1) in the quotient above the 6 (tens place). Write the product (12) below the 16.

$$\begin{array}{r} \text{x 1} \\ 12\overline{)168} \\ -12 \end{array}$$

Subtract: 16 – 12 = 4. Write this difference under the 12.

$$\begin{array}{r} \text{x 1} \\ 12\overline{)168} \\ -12 \\ \hline 4 \end{array}$$

Drop Down: Bring the next number (8) in the dividend down to the right of the partial difference in the ones column.

$$\begin{array}{r} \text{x 1} \\ 12\overline{)168} \\ -12\downarrow \\ \hline 4\ 8 \end{array}$$

Repeat:
1. Divide 48 by 12.

2. Multiply 12 × <u>4</u> = 48. The factor (4) goes in the quotient above the 8 (ones column).

$$\begin{array}{r} \text{x 1 4} \\ 12\overline{)168} \\ -12\downarrow \\ \hline 4\ 8 \\ -4\ 8 \\ \hline 0 \end{array}$$

3. Subtract the product from 48. (48 − 48 = 0)

4. Drop Down: Nothing to drop down from dividend.

Always check your answer using multiplication.
If 168 ÷ 12 = 14, then 14 × 12 = 168.

168 ÷ 12 = 14

EXAMPLE 2

A bolt of fabric measured 1,380 inches in length. The seamstress wants to cut it into 16 equal pieces for her students. How long will each piece be?

SOLUTION

Divide: 1 cannot be divided by 16. Put an X over the 1 in the quotient. 13 cannot be divided by 16. Put an X over the 3 in the quotient. 138 can be divided by 16.

$$\begin{array}{r} \text{x x} \\ 16\overline{)1380} \end{array}$$

Multiply: $16 \times \underline{8} = 128$ (closest number to 138). Write the factor (8) in the quotient above the 8 (tens place). Write the product (128) below the 138.

$$\begin{array}{r} \text{x x 8} \\ 16\overline{)1380} \\ -128 \end{array}$$

Subtract: $138 - 128 = 10$. Write this difference under the 138.

$$\begin{array}{r} \text{x x 8} \\ 16\overline{)1380} \\ -128 \\ \hline 10 \end{array}$$

Drop Down: Bring the next number (0) in the dividend down to the right of the partial difference in the ones column.

$$\begin{array}{r} \text{x x 8} \\ 16\overline{)1380} \\ -128\downarrow \\ \hline 100 \end{array}$$

Repeat:
1. Divide 100 by 16.

2. Multiply $16 \times \underline{6} = 96$. The factor (6) goes in the quotient above the 0 (ones column).

3. Subtract the answer from 100. $(100 - 96 = 4)$

$$\begin{array}{r} \text{x x 86} \quad \frac{4}{16} \text{ or } 86\frac{1}{4} \\ 16\overline{)1380} \\ -128 \\ \hline 100 \\ -96 \\ \hline 4 \end{array}$$

4. Drop Down: Nothing to drop down from dividend. Write the remainder as a fraction after the whole number part of the quotient.

$$1380 \div 16 = 86\frac{4}{16} = 86\frac{1}{4}$$

The seamstress cut each piece $86\frac{1}{4}$ inches long.

> Always check your answer using multiplication.
>
> If $1380 \div 16 = 86\frac{1}{4}$, then $86 \times 16 + 4 = 1380$.

> Writing the remainder as $\frac{4}{16}$ (simplified as $\frac{1}{4}$) in this case makes sense because this is a measurement. The seamstress could use a fraction of the material for each piece.

EXERCISES

Find each quotient. Write any remainder as R__.

1. $392 \div 14$

2. $16\overline{)576}$

3. $365 \div 12$

4. $17\overline{)384}$

5. $25\overline{)778}$

6. $498 \div 24$

Find each quotient. Write any remainder as a fraction.

7. $20\overline{)410}$

8. $438 \div 18$

9. $910 \div 42$

10. $524 \div 16$

11. $35\overline{)749}$

12. $12\overline{)746}$

Find each quotient. Label your answer.

13. Chad delivered 297 gallons of drinking water to companies over 18 days. If he delivered the same number of gallons each day, how many gallons did he deliver each day?

14. Scott planted blueberry bushes on his farm. He had 456 bushes. He planted 30 blueberry bushes in each row. How many rows did he plant? How many blueberry bushes were left?

15. Creamilicious Ice Cream Company can put 12 bars (any flavor) in each box they sell. For each flavor below, how many boxes can they fill? How many are left over?
  a. Cookie Ice Cream Sandwich
  b. Strawberry Freeze Bar
  c. Double Chocolate Bar
  d. Caramel Fudge Bar

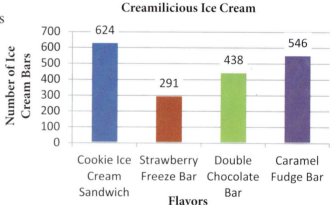

Find each quotient. Write any remainder as R__ and as a fraction.

16. $11\overline{)1244}$

17. $2812 \div 13$

18. $1342 \div 12$

19. $15\overline{)3022}$

20. $14\overline{)2621}$

21. $4384 \div 15$

22. What are the missing numbers in the dividend $2\square8\square$? Show how you know.

$$
\begin{array}{r}
213 \text{ R}1 \\
14\overline{)2\square8\square} \\
-28 \\
\hline
18 \\
-14 \\
\hline
4\square \\
-42 \\
\hline
1
\end{array}
$$

Find the value of each expression.

23. 9.2 – 4
24. 49.82 – 8.9
25. 16.031 – 7.825

26. Marliese took 325 pounds of produce to 6 produce stands. Each stand received the same number of pounds of produce. How many pounds did each produce stand receive?

27. Trent set the tables for a fundraiser meal. There were 215 plates to set on tables. Trent could place 8 plates on each table. How many tables did he completely set? How many plates were left over.

TIC-TAC-TOE ~ DRAWING DIMENSIONS

Measure two rooms in your house or school using metric measurements (centimeters to the nearest centimeter).

Example: Length of one room is 381 centimeters

Using grid paper, set up a scale where each side on a grid square represents 20 centimeters. Divide each of your measurements by 20. This will tell you how many squares to use when drawing your rooms to scale. Example: 381 centimeters ÷ 20 = 19.05. You will trace a bit more than 19 square lengths to show this on your grid paper. Draw each room to scale.

TIC-TAC-TOE ~ RACING THE CLOCK

Measure out a short distance to run in meters (less than 10 meters).

Multiply that distance by 4.

Using a stopwatch, time 5 friends as they run the short distance. Record their times.

Using a stopwatch, time the same 5 friends as they run the longer distance. Record their times.

Divide each friend's longer distance time by 4. Is the time equal to, less than or greater than their time for the shorter distance? Why or why not?

Make a poster to display the times for each race and your findings.

DIVIDING DECIMALS BY WHOLE NUMBERS

LESSON 2.5

 Find quotients of expressions where decimals are divided by whole numbers.

Kainan's basketball coach purchased water bottles for the twelve team members. The total cost was $4.80. The cost was divided equally among the players. How much did Kainan owe?

When you have a full price and need to find out the unit price, you need to divide. Use base-ten blocks or money to visualize dividing decimals.

To find out how much Kainan owed for his water bottle, he divided $4.80 by 12.

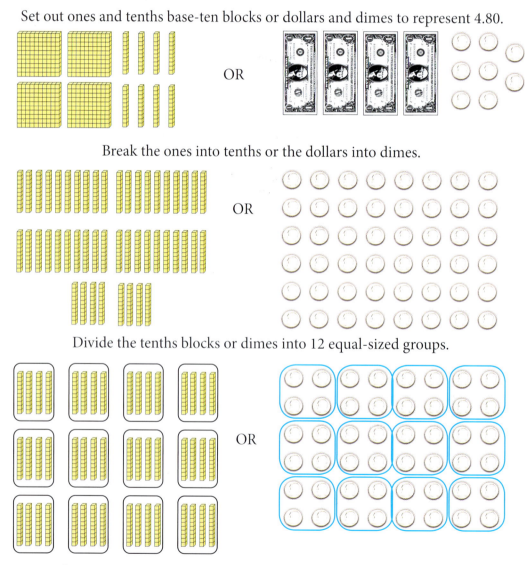

Set out ones and tenths base-ten blocks or dollars and dimes to represent 4.80.

OR

Break the ones into tenths or the dollars into dimes.

OR

Divide the tenths blocks or dimes into 12 equal-sized groups.

OR

Each group represents four tenths or 0.4 or $0.40. Kainan owes $0.40 for the water bottles.

DIVIDING DECIMALS BY WHOLE NUMBERS

1. Divide as if both dividend and divisor were whole numbers.
2. Move the decimal point into the quotient directly above the decimal point in the dividend.
3. Insert zeros if necessary to hold place value.
4. Some answers will need to be rounded to a given place value.

EXAMPLE 1

Joe bought 3.9 pounds of candy corn for a harvest party. He divided it among himself and nine friends. How many pounds of candy corn did each person get?

SOLUTION

Write the problem.

$$3.9 \div 10$$

Divide as if the divisor and dividend were whole numbers.

```
        39
  10 ) 3.90
      −30
        9 0
       −9 0
          0
```

> Insert a zero on the end of the dividend. Continue dividing.

> When subtracting these numbers ignore the decimal point.

Move the decimal point into the quotient directly above the decimal point in the dividend.

```
        .39
  10 ) 3.90
```

Each person received 0.39 pounds of candy corn.

EXAMPLE 2

Find the value of 2.97 ÷ 30.

SOLUTION

Divide as if the divisor and dividend were whole numbers.

```
         99
  30 ) 2.970
      −270
        270
       −270
          0
```

> Insert a zero on the end of the dividend. Continue dividing.

Move the decimal point into the quotient directly above the decimal point in the dividend.

```
        ._99
  30 ) 2.970
```

Insert a zero to hold the tenths place.

```
        .099
```

$$2.97 \div 30 = 0.099$$

Sometimes you may need to round to a designated place-value. Divide until the quotient has one more place-value position than where it is being rounded to. This helps you know whether to round up or down. You may need to add one or more zeros at the end of the decimal dividend to find the quotient.

EXAMPLE 3

Janease spent \$3.22 to download three songs. Each song cost about the same amount. How much did she spend on each song? Round to the nearest penny (hundredth) if necessary.

SOLUTION

Write the problem.

$$3.22 \div 3$$

Divide as if the divisor and dividend were whole numbers. Continue until you have gone one place value past the hundredth position.

```
        1073
  3 )3.220
     -3
      02
     -00
      22
     -21
      10
      -9
       1
```

Insert a zero on the end of the dividend and continue dividing.

Move the decimal point into the quotient directly above the decimal point in the dividend.

```
       1.073
  3 )3.220
```

Round to the nearest penny.

1.07

Janease spent about \$1.07 on each song.

EXAMPLE 4

Find the value of 16 ÷ 5.

SOLUTION

Place a decimal point in the dividend at the end of the whole number.

```
  5 )16.
```

Divide.

```
      32
  5 )16.0
     -15
      10
     -10
       0
```

Insert a zero on the end of the dividend. Continue dividing.

Move the decimal into the quotient.

```
      3.2
  5 )16.
```

16 ÷ 5 = 3.2

EXERCISES

Find each quotient.

1. 3.8 ÷ 2

2. $21.30 ÷ 3

3. 36.5 ÷ 5

4. $15.36 ÷ 6

5. 11.2 ÷ 4

6. 14.32 ÷ 8

7. The grocer charged Tory $4.47 for grapes. She bought three pounds of grapes. How much did each pound of grapes cost?

8. Carole bought six bouquets of flowers. Each bouquet cost the same amount. She spent a total of $59.88. How much did each bouquet cost?

9. Vashti was paid for every weed she pulled. She pulled 25 weeds. She earned $3.00. How much was she paid for each weed?

10. 0.63 ÷ 7

11. 1.05 ÷ 12

12. 0.97 ÷ 10

13. 1.47 ÷ 15

14. 1.68 ÷ 21

15. 4.6 ÷ 4

16. There were 0.5 gallons of milk in the refrigerator for a family of eight. Each person drank the same amount of milk. How much of a gallon did each person drink?

17. Twelve people split $18.00 equally. How much money did each person receive?

Find each quotient. Round your answer to the nearest hundredth.

18. $13.42 ÷ 6

19. 8.7 ÷ 7

20. 41 ÷ 7

21. 11.9 ÷ 3

22. $50.63 ÷ 5

23. 21.493 ÷ 4

Store	Weight in pounds	Total Price
Alan's	4	$6.60
Food Market	2	$3.55
Save More	3	$5.10

24. The table gives information for purchasing dried apricots at three different grocery stores.
   a. Find the price per pound at each store. Round to the nearest penny.
   b. Which store has the best buy?

REVIEW

25. Estimate each product.
   a. 7.8 × 2.1
   b. 72.4 × 12.6

26. Find each product
   a. 7.8 × 2.1
   b. 72.4 × 12.6

27. Estimate each product.
   a. 22.6 × 9
   b. 17.45 × 14.82

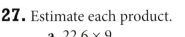

28. Find each product.
   a. 22.6 × 9
   b. 17.45 × 14.82

DIVIDING DECIMALS BY DECIMALS

LESSON 2.6

 Find quotients of expressions where decimals are divided by decimals.

Dividing requires you to look at how many groups of one number fit into another number. When dividing decimals, the same approach applies.

24 ÷ 6 can be read as, "How many groups of 6 fit in the number 24?"
2.4 ÷ 0.6 can be read as, "How many groups of 0.6 fit in the number 2.4?"

Base-ten blocks can be used to model 2.4 ÷ 0.6.

Set out base-ten blocks to model the dividend, 2.4.

Substitute any ones blocks with tenths because the divisor is tenths.

Move the tenths into groups the size of the divisor.

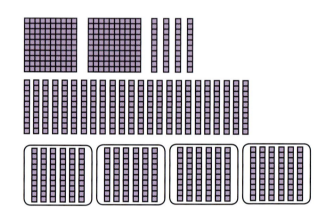

There are 4 groups of 0.6 in 2.4.   2.4 ÷ 0.6 = 4

This means that 2.4 is four times greater than 0.6.   0.6 × 4 = 2.4

DIVIDING DECIMALS BY DECIMALS

1. Change the divisor into a whole number by moving the decimal point to the right.
2. Move the decimal point in the dividend the same number of places to the right as it was moved in the divisor.
3. Divide the dividend by the divisor.
4. Move the decimal point into the quotient directly above the decimal point in the dividend.
5. Insert zeros, if necessary, to hold place values.

---

58 *Lesson 2.6 ~ Dividing Decimals by Decimals*

EXAMPLE 1	**Thuyet purchased 3.2 pounds of fertilizer for $4.48. To determine how much he paid per pound you must find $4.48 ÷ 3.2.**

SOLUTION

Rewrite the problem. When dividing by a decimal the divisor needs to be a whole number. Do this by moving the decimal point to the right in the divisor first.

Divisor

3.2.

The decimal point in the dividend must be moved the same number of places to the right as in the divisor.

Divisor

3.2.

32

> The decimal point was moved one place to the right.

Dividend

4.4.8

44.8

> Move the decimal point one place to the right.

Find the quotient using the new dividend and divisor.

44.8 ÷ 32

$$\begin{array}{r} 1.4 \\ 32\overline{)44.8} \\ -32 \\ \hline 128 \\ -128 \\ \hline 0 \end{array}$$

Thuyet's fertilizer cost $1.40 per pound.

EXAMPLE 2	**Find the value of 67.2 ÷ 0.56.**

SOLUTION

Change the divisor into a whole number and move the decimal point the same number of places to the right in both the divisor and the dividend.

$$0.56\overline{)67.20}$$

> Insert a zero on the end of the dividend to hold place-value when the decimal point is moved.

Divide the dividend by the divisor as if both were whole numbers.

$$\begin{array}{r} 120 \\ 56\overline{)6720} \\ -56 \\ \hline 112 \\ -112 \\ \hline 00 \end{array}$$

> Zero is the last digit in the quotient because 0 ÷ 56 = 0

Move the decimal point into the quotient directly above the decimal point in the dividend.

$$\begin{array}{r} 120. \\ 56\overline{)6720.} \end{array}$$

67.2 ÷ 0.56 = 120

EXAMPLE 3

Joe makes $8.30 per hour. Macie makes $12.45 per hour. How many times greater is Macie's hourly pay than Joe's hourly pay?

SOLUTION

To find how many times greater Macie's hourly pay is, write a multiplication equation.

$8.30 × _____ = $12.45

Use division to find the missing number.

$12.45 ÷ $8.30

Change the divisor into a whole number. Move the decimal point the same number of places to the right in both the divisor and the dividend.

8.3⌒|12.45⌒

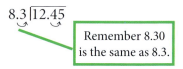
Remember 8.30 is the same as 8.3.

Divide the dividend by the divisor as if both were whole numbers.

$$\begin{array}{r} 15 \\ 83\overline{)124.5} \\ -\underline{83} \\ 415 \\ -\underline{415} \\ 0 \end{array}$$

Move the decimal point into the quotient directly above the decimal point in the dividend.

$$\begin{array}{r} 1.5 \\ 83\overline{)124.5} \end{array}$$

Macie makes 1.5 times the amount per hour that Joe makes.

EXAMPLE 4

Landon put 10.912 gallons of gas into his car last week. This week he put in 10.02 gallons of gas. How many times greater was the amount of gasoline last week than this week? Round to the nearest tenth.

SOLUTION

Change the divisor into a whole number. Move the decimal point the same number of places to the right in both the divisor and the dividend.

10.02⌒|10.912⌒

Divide the dividend by the divisor as if both were whole numbers.

$$\begin{array}{r} 108 \\ 1002\overline{)1091.20} \\ -\underline{1002} \\ 892 \\ -\underline{0} \\ 8920 \\ -\underline{8016} \\ 904 \end{array}$$

Stop dividing when you have one more place-value position in the quotient than needed.

Move the decimal point into the quotient directly above the decimal point in the dividend.

$$\begin{array}{r} 1.08 \\ 1002\overline{)1091.20} \end{array}$$

Round to the nearest tenth.

1.08 ≈ 1.1

Last week's fill-up was approximately 1.1 times more than this week's fill-up.

EXERCISES

Rewrite each division problem so the divisor is a whole number.

1. 4.8 ÷ 0.8

2. 10.28 ÷ 2.55

3. 124 ÷ 3.4

4. 10.85 ÷ 7.75

5. 18.29 ÷ 3.1

6. 9.9 ÷ 6.62

Find each quotient.

7. 2.7 ÷ 0.3

8. 4.8 ÷ 3.2

9. 4.9 ÷ 0.07

10. 9.72 ÷ 5.4

11. 14 ÷ 0.4

12. 0.115 ÷ 0.025

13. 32.8 ÷ 0.5

14. 7.1 ÷ 2.84

15. 4.42 ÷ 2.6

16. Susan spent $71.50 at the mall. Elise spent $28.60. How many times greater was Susan's spending than Elise's spending?

17. Sean's pumpkin weighed 33.12 pounds. Kevin's pumpkin weighed 9.6 pounds. How many times heavier was Sean's pumpkin than Kevin's pumpkin?

Find each quotient. Round to the nearest tenth, if necessary.

18. 12.9 ÷ 4.6

19. 11.8 ÷ 2.7

20. 1.287 ÷ 0.25

21. 15.95 ÷ 3.95

22. 21.90 ÷ 6.89

23. 10.1 ÷ 1.01

24. The state of Oregon had an approximate population of 3.7 million people in 2006. The state of Washington had an approximate population of 6.4 million people in the same year. How many times greater was the population of Washington than the population of Oregon?

25. Yuki bought 1.6 pounds of lettuce for $2.05. What does the lettuce cost per pound, to the nearest cent?

26. The state of Michigan had a population of about 10 million people in 2008. The State of Indiana had a population of about 6.38 million in the same year. How many times greater was the population of Michigan than the population of Indiana?

27. Sharla weighs 52.4 kilograms. Troy weighs 83.84 kilograms. Troy's weight is how many times greater than Sharla's weight?

Find each product. Round to the nearest hundredth, if necessary.

28. 3.1×3.1

29. 9.52×4.5

30. 10.7×5.39

31. 2.8×0.072

32. 0.18×1.5

33. 15.1×0.48

TIC-TAC-TOE ~ DECIMAL POETRY

Ideas and thoughts can be expressed in poetry as a written form of art. Choose two different forms of poetry. Write one poem about multiplying decimals by decimals. Write another poem about dividing decimals by decimals.

Nonet

A nonet is a poem that consists of nine lines. Rhyming is optional. The first line has 9 syllables, the second line has 8 syllables. Continue through the ninth line which has only one syllable.

Free Verse

Free Verse is poetry that doesn't follow traditional poetry rules (meter and rhyme). Line breaks are used to create meaning and allow the reader to slow down, stop or speed up their reading to accentuate a part of the poem.

Song

A song is a rhythmic poem that has verses and a chorus. The chorus is repeated between each verse.

TIC-TAC-TOE ~ TEACHER, TEACHER

Make a teacher's manual for teaching the concept of dividing decimals. Explain the process in your own words. Give examples of different types of decimal division problems that a teacher could use when teaching his or her class. Work each problem out to show the steps.

Have a parent or classmate follow your process to see if it works. Make changes to your directions so they are clear and useful, if needed.

Vocabulary

algorithm
dividend
divisor

factors
fraction

product
quotient
remainder

 Find products of expressions involving multi-digit whole numbers.
Find products of expressions involving decimals.
Find quotients of expressions where whole numbers are divided by 1-digit whole numbers, including remainders.
Find quotients of expressions where whole numbers are divided by multi-digit whole numbers, including remainders.
Find quotients of expressions where decimals are divided by whole numbers.
Find quotients of expressions where decimals are divided by decimals.

Lesson 2.1 ~ Multiplying by 2-Digit Numbers

Find each product.

1. 35×8

2. 23×6

3. 82×7

4. 12×15

5. 37×21

6. 53×19

7. 17×32

8. 26×45

9. 67×39

10. Vaughn uses 42 feet of thick rope for the ties on each hammock he makes. He made 7 hammocks last month. How many feet of rope did he use?

11. Ahmei read 64 pages of her book each day for 12 days. How many pages did she read in all?

Lesson 2.2 ~ Multiplying Decimals

Find each product.

12. 7.3×4

13. 3×6.21

14. 5×11.05

15. 3.4×2.6

16. 6.23×4.7

17. 7.14×5.23

18. 0.048×5

19. 8.9×0.039

20. 9.64×0.98

21. Alejandro bought 5.37 pounds of oranges. They cost $0.49 per pound. How much did Alejandro spend on oranges? Round to the nearest penny.

22. Lucia bought games for her video game system. The games cost $19.99 each. How much money did Lucia need for three games?

Lesson 2.3 ~ Dividing by 1-Digit Numbers

• •

Find each quotient. Write any remainder as R ____.

23. 75 ÷ 3

24. 53 ÷ 4

25. 82 ÷ 7

26. 634 ÷ 6

27. 299 ÷ 5

28. 436 ÷ 8

29. Raul and Miguel came up with different answers for the same equation.
  a. Who is correct?
  b. Describe the error the other person made.

Raul's Work	Miguel's Work
$\begin{array}{r} 145 \text{ R3} \\ 6\overline{)873} \\ -6 \\ \hline 27 \\ -24 \\ \hline 33 \\ -30 \\ \hline 3 \end{array}$	$\begin{array}{r} 145 \frac{3}{873} \\ 6\overline{)873} \\ -6 \\ \hline 27 \\ -24 \\ \hline 33 \\ -30 \\ \hline 3 \end{array}$

Lesson 2.4 ~ Dividing by 2-Digit Numbers

• •

Find each quotient. Write any remainder as R____.

30. 543 ÷ 12

31. 764 ÷ 24

32. 658 ÷ 18

33. 1323 ÷ 11

34. 4652 ÷ 30

35. 3980 ÷ 26

36. Tom edited books. He edited 605 pages in 30 days. Assume he edited the same number of pages each day. How many pages did he edit each day?

37. Marissa made flower arrangements for a reception. She had 408 flowers. There were 25 tables that needed arrangements. Every table had the same number of flowers in the arrangement. How many flowers were in each arrangement? How many were left over?

38. Jacques rode the train 749 miles. The train made 28 stops. If the stops were exactly the same distance apart, how many miles were there between stops?

Find each quotient. Round to the nearest hundredth, if necessary.

39. $9.9 \div 2$

40. $13.53 \div 3$

41. $32.5 \div 5$

42. $2.7 \div 9$

43. $0.88 \div 7$

44. $0.55 \div 10$

45. $19.33 \div 6$

46. $54.54 \div 6$

47. $38.675 \div 7$

48. Aubrey had 3.5 pounds of ground beef for tacos. She made 25 tacos. How much ground beef did she put into each taco?

Lesson 2.6 ~ Dividing Decimals by Decimals

Find each quotient. Round to the nearest hundredth, if necessary.

49. $6.7 \div 0.3$

50. $14.04 \div 3.2$

51. $16.4 \div 2.35$

52. $50.5 \div 0.58$

53. $28.92 \div 17.4$

54. $33.76 \div 6.56$

55. Kaori spent \$99.55 for her cell phone. Yin spent \$258.83 to purchase her cell phone. How many times greater was the cost of Yin's cell phone than Kaori's cell phone?

56. Elsa made 3.5 cups of guacamole. A serving of guacamole was 0.125 cup. How many servings did Elsa make?

Tic-Tac-Toe ~ Color By Numbers

Draw an outline of a picture or use a picture from a coloring book.

Decide the colors you want for the picture (6 or more).

Write multiplication or division problems that use decimals and equal 6 or more different answers (the number of colors you use is the number of answers you need).

Divide picture into sections if there are not already given sections.

Set up answer key (Answer = _____ color)

Put the multiplication or division problems in the spots designed for the answer that gives you the color you want.

**JOE
FARMER**

I am a farmer. Farming involves planting, irrigating, harvesting and storing of the food that we eat. This includes fertilizing and cultivating the fields where the crops are grown. Today's American farmer uses a wide variety of equipment and technology to produce the safest and highest quality food stuffs in the world. Nearly everything you eat or drink has been produced by a proud farmer.

I would have a difficult time farming if I didn't have a basic understanding of mathematics. I use math to calculate the area of the fields I am planting. I also use math to determine how much seed, fertilizer and herbicides to buy. I must also estimate how much water to use, as well as how much moisture is in the ground and crops that I harvest. I use many conversion factors and algebraic equations to accomplish this. I also use math to prepare budgets and financial statements so I can qualify for the financing necessary to operate my farm.

Some people working on farms do not have very much education at all. Due to the shortage of farm labor, many entry level farm jobs are filled by workers without much education. However, a person who desires to manage or own a farm should at least have a high school education. Many farmers obtain bachelor's degrees from college in either agriculture or a related field.

People just starting to work on a farm as farm laborers often make only minimum wage. As they gain more experience they can make as much as $12 per hour. Farm managers make an average of $55,000 per year, with the highest paid managers making as much as $90,000 per year.

Farming is as much a lifestyle as an occupation. There are many challenges and much hard work. However, there is also a great deal of satisfaction. Farming has given me the opportunity to have my own business producing something that people really need. The best part of farming is being outdoors working side by side with my family, doing what we love.

CORE FOCUS ON DECIMALS & FRACTIONS
BLOCK 3 ~ UNDERSTANDING FRACTIONS

LESSON 3.1   GREATEST COMMON FACTOR --- 69
              EXPLORE! UNIVERSITY SALES

LESSON 3.2   EQUIVALENT FRACTIONS -- 74
              EXPLORE! CREATING EQUIVALENT FRACTIONS

LESSON 3.3   SIMPLIFYING FRACTIONS --- 79
              EXPLORE! FRACTION HOMEWORK

LESSON 3.4   LEAST COMMON MULTIPLE --- 83

LESSON 3.5   ORDERING AND COMPARING FRACTIONS -- 87
              EXPLORE! WHICH IS LARGER?

LESSON 3.6   MIXED NUMBERS AND IMPROPER FRACTIONS --- 92
              EXPLORE! CHOCOLATE CHIP COOKIES

LESSON 3.7   MEASURING IN INCHES --- 96
              EXPLORE! USING A CUSTOMARY RULER

REVIEW   BLOCK 3 ~ UNDERSTANDING FRACTIONS --- 101

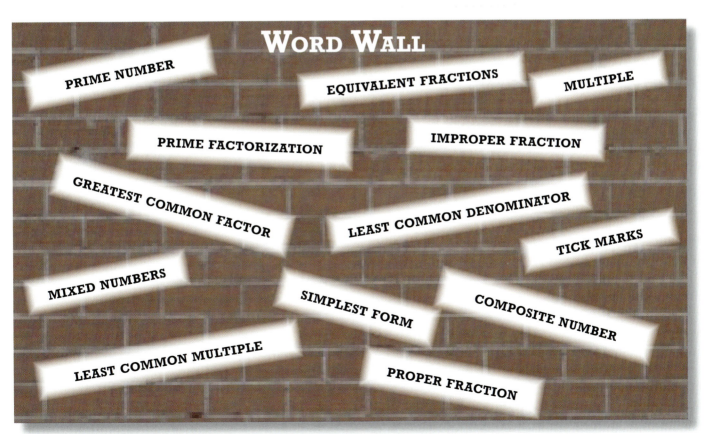

WORD WALL

PRIME NUMBER

EQUIVALENT FRACTIONS

MULTIPLE

PRIME FACTORIZATION

IMPROPER FRACTION

GREATEST COMMON FACTOR

LEAST COMMON DENOMINATOR

TICK MARKS

MIXED NUMBERS

SIMPLEST FORM

COMPOSITE NUMBER

LEAST COMMON MULTIPLE

PROPER FRACTION

BLOCK 3 ~ UNDERSTANDING FRACTIONS
TIC-TAC-TOE

EXPLORE PRIME

Investigate prime numbers. Decide if there is a pattern for finding prime numbers.

See page 73 for details.

CONCAVE AND CONVEX

Create concave and convex figures using seven different line segments.

See page 105 for details.

FAVORITES

Survey at least 50 people. Write results as fractions. Create a graph to display your results.

See page 91 for details.

OCCUPATIONS

Research and write about an occupation that uses fractions on a daily basis.

See page 73 for details.

EDIBLE FRACTIONS

Use a bag of multicolored candies to write, simplify and order fractions.

See page 86 for details.

FRACTION GAME

Make a game using equivalent fractions.

See page 78 for details.

EQUIVALENT FRACTIONS

Make a poster to display sets of equivalent fractions.

See page 78 for details.

COINS AND FRACTIONS

Create a table that displays the worth of coins as fractions of a dollar.

See page 100 for details.

RECIPE MIX-UP

Create a Recipe Mix-Up cookbook using improper and equivalent fractions.

See page 95 for details.

GREATEST COMMON FACTOR

 Find the greatest common factor (GCF) of a set of numbers.

Factors are numbers that can be multiplied together to find a product. For example, 2 and 4 are factors of 8 because $2 \times 4 = 8$.

When a whole number has only two possible factors (1 and the number itself), it is called a **prime number**. A whole number larger than one is called a **composite number** when it has more than two factors.

EXAMPLE 1

Determine if 12 is a prime or composite number.

SOLUTION

List the pairs of numbers that have a product of 12. These are the factors of 12.

$$1 \times 12 \qquad 12 \times 1$$
$$2 \times 6 \qquad 6 \times 2$$
$$3 \times 4 \qquad 4 \times 3$$

These are the same as the first column but in reverse order.

List each factor once, even if it is repeated.

The factors of 12 are 1, 2, 3, 4, 6 and 12.

There are more than two factors so the number 12 is composite.

The **greatest common factor (GCF)** of two or more numbers is the greatest factor that is a whole number common to all the numbers. The greatest common factor can be used to solve problems involving real-life situations.

EXPLORE!

UNIVERSITY SALES

Bracken had 36 University of Miami shirts and 42 Florida State University shirts to sell. He wants to stack them in piles that would all have the same number of shirts. He does not want to mix the two types of shirts. What is the greatest number of shirts that can be stacked in each pile? Find the GCF of 36 and 42.

Step 1: Find all factors of 36 by filling the boxes with the missing factors. Make a list of all factors of 36.

$$\Box \times 36 \qquad 2 \times \Box \qquad \Box \times 12 \qquad 4 \times \Box \qquad \Box \times 6$$

Step 2: Find all factors of 42 by filling the boxes with the missing factors. Make a list of all factors of 42.

$$1 \times \Box \qquad 2 \times \Box \qquad 3 \times \Box \qquad \Box \times 7$$

Step 3: Circle the common factors. Common factors are factors that are the same for both 36 and 42.

Step 4: Draw a Venn diagram like the one to the right on a sheet of paper. Write "Factors of 36" on the outside of the left circle and "Factors of 42" on the outside of the right circle.

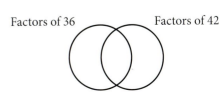

Factors of 36   Factors of 42

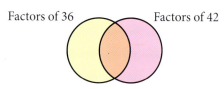

Step 5: Place all factors on the Venn diagram. The factors that both numbers have in common go in the overlapping part of the circles. The remaining factors of 36 go in the left circle. The remaining factors of 42 go in the right circle.

Step 6: Look at the common factors where the circles overlap. Circle the largest number. This is the greatest common factor (GCF).

Step 7: Use the GCF to answer the question in the problem at the beginning of the Explore! in a complete sentence.

Step 8: Repeat **Steps 1-6** to find the GCF of the following pairs of numbers:
   a. 15 and 25   **b.** 18 and 30   **c.** 24 and 40

There are other methods to find the greatest common factor. **Prime factorization** is shown when any composite number is written as a product of all its prime factors.

EXAMPLE 2

Two local teams went to soccer camp together. At the camp the teams were asked to split into equal amounts for cabin groups. The players did not want to room with players from other teams. The camp directors want the largest number possible in each cabin. How many players will be in each cabin?

Team 1	36 players
Team 2	30 players

SOLUTION

Use prime factors to find the GCF. Prime factors are factors that are prime numbers.

Write each number as products of two factors.

Continue to write each number as products of two factors until only factors that are prime numbers remain.

Write the factors out for each number. This is called the prime factorization. Highlight the common prime factors.

$$36 = 2 \times 2 \times 3 \times 3 \qquad 30 = 2 \times 3 \times 5$$

Find the product of the common prime factors. This is the GCF.

$$GCF = 2 \times 3 = 6. \text{ The GCF is } 6.$$

Six players will be in each cabin.

EXAMPLE 3

Reagan Middle School students were asked to sit in equal rows for the assembly. There were 98 sixth graders, 84 seventh graders and 112 eighth graders. The teachers did not want grade levels sitting together, but the rows were to be as wide as possible. How many students should sit in each row?

SOLUTION

List the factors of each number. Highlight the common factors.

Factors of 98: **1**, **2**, **7**, **14**, 49, 98
Factors of 84: **1**, **2**, 3, 4, 6, **7**, 12, **14**, 21, 28, 42, 84
Factors of 112: **1**, **2**, 4, **7**, 8, **14**, 16, 28, 56, 112

Find the GCF.

The GCF of 98, 84 and 112 is 14.

Fourteen students should sit in each row.

FIND THE GREATEST COMMON FACTOR

1. List the factors for each number.
2. Highlight the common factors.
3. Identify the GCF (greatest common factor).

EXERCISES

List the factors of each number. State whether each number is prime or composite.

1. 4

2. 3

3. 8

4. 7

5. 14

6. 20

7. 29

8. 16

9. 27

10. Alexis had 8 bracelets and 10 necklaces. She wanted to put equal amounts of each item into small travel containers. She did not want to mix the bracelets and necklaces. What is the largest number of items she could put in a travel container?

a. Draw a Venn diagram. Put the number "8" on one side and "10" on the other side.

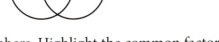

b. List all factors for both numbers. Highlight the common factors.
c. Place all uncommon factors for the 1ˢᵗ number in the left circle.
d. Place all uncommon factors for the 2ⁿᵈ number in the right circle.
e. Place the factors both numbers have in common in the overlapping part of the circles.
f. What is the GCF?

Copy the following Venn diagrams. Write the factors on the Venn diagrams. Find the greatest common factor for each set of numbers.

11. 9   12
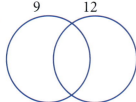

12. 15   20
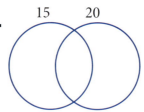

13. 54   64
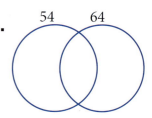

List the factors for each number. Circle the greatest common factor for each set.

14. 3 and 6

15. 6 and 9

16. 12 and 18

17. 32, 48 and 64

18. 30, 45 and 60

19. 24, 40 and 54

Use prime factorization to find the greatest common factor of each set of numbers.

20. Tyler had 45 baseball cards and 54 basketball cards. He organized them into equal rows for a display. He did not want to mix the basketball and baseball cards in a row. What was the greatest number of cards in each row?

    a. Write each number as the product of any two of its factors.

    b. Continue to write each factor as the product of two factors until only prime factors remain.

    c. Write the prime factorization for each original number. Highlight the common prime factors.

    d. Find the product of the common prime factors.

    e. What is the GCF?

21. 18 and 24

22. 64 and 80

23. 84 and 56

24. Two sixth grade classes went to the local theater to watch a movie. They reserved seats ahead of time so each class could sit together in equal rows. One class had 28 students and the other had 21 students. What is the greatest number of students that could sit in each row?

25. Camilla separated prizes for games at the carnival. She had 72 Choco Bars and 90 Peanut Blitzes. She put the most candy bars possible in each bag without mixing the two types. Each bag needed an equal amount of candy. How many candy bars did she put in each bag?

26. This lesson contained three methods to find the greatest common factor: Venn diagrams, lists and prime factorization. Which method do you like best? Why?

Find each quotient. Round to the nearest hundredth, if necessary.

27. 6.49 ÷ 4

28. 31.67 ÷ 6

29. 102.987 ÷ 5

30. 23.35 ÷ 2.1

31. 39.87 ÷ 3.2

32. 987.34 ÷ 10.4

33. 825.25 ÷ 0.25

34. 468.441 ÷ 5.65

35. 225.92 ÷ 1.32

TIC-TAC-TOE ~ EXPLORE PRIME

A prime number is a whole number larger than 1 that is the product of only two factors, 1 and itself. All other numbers are called composite numbers.

1. Copy the chart. Fill in the numbers from 2 through 50. Explain how you found the prime numbers. Give examples of at least 3 factors for each composite number.

Prime	Composite
2 = 2 × 1	4 = 4 × 1 or 2 × 2
3 = 3 × 1	6 = 6 × 1 or 2 × 3

2. Is there a pattern for finding prime numbers? Research prime numbers on the internet or in books. Write on one researcher's ideas about patterns with prime numbers. Include where you found the information (the exact web-site, book, etc.) in your paper.

TIC-TAC-TOE ~ OCCUPATIONS

Many people use fractions in their daily work. Look at different occupations where people use fractions in their jobs. Pick one occupation to research.

Here are some things to look for.
- **a.** What do people in that occupation do daily?
- **b.** Why is that occupation necessary in our world?
- **c.** How do people in that occupation use fractions?

Create a magazine spread (1-2 pages) with your research.

1. Write an article that includes the important information from your research. Be sure to include answers to the three questions above.

2. Choose to take, print or draw one or more pictures for the spread.

Possible article ideas:
- Write the article as an interview between you (the magazine publisher) and a person in that occupation.
- Write the article as your view of "a day in the life of a person who works in…"
- Create a fictional person who is in that occupation. Write the article from his/her viewpoint.
- Write letters from fictional people in that occupation to your magazine explaining different aspects of their job.

EQUIVALENT FRACTIONS

 Write equivalent fractions.

A recent survey asked girls and boys how they spend their free time. Three out of six girls said they spend their free time playing sports. One-half of the boys said they spend their free time playing sports. The results of the survey can be written using fractions.

A fraction is a number written as $\frac{\text{numerator}}{\text{denominator}}$.

A fraction represents part of a whole number. The denominator of a fraction cannot be 0. The linc between the numerator and denominator can be read "out of". In the survey above, $\frac{3}{6}$ of the girls and $\frac{1}{2}$ of the boys said they spend their free time playing sports.

There are times when information needs to be compared to make it easier to understand. Equivalent fractions allow you to determine whether two fractions are equal. **Equivalent fractions** are two fractions that name the same amount.

To compare the fraction of girls that play sports to the fraction of boys that play sports, you can use models.

Draw two equal-sized rectangles.
Each rectangle represents one whole unit.

Divide each rectangle into equal-sized sections using the denominator to determine the number of sections.

$\frac{3}{6}$

$\frac{1}{2}$

Use the value of the numerator to determine the number of sections to color in for each rectangle.

$\frac{3}{6}$

$\frac{1}{2}$

The same amount of each rectangle is colored. This means that $\frac{3}{6}$ is equivalent to $\frac{1}{2}$. According to the survey, the same portion of girls play sports in their free time as boys.

CREATING EQUIVALENT FRACTIONS

Equivalent fractions are created by multiplying or dividing the numerator and denominator by the same amount.

EXAMPLE 1

EXAMPLE 1 | Use multiplication to find the missing number in the equivalent fractions.

a. $\frac{1}{2} = \frac{\square}{8}$　　　　　b. $\frac{3}{8} = \frac{6}{\square}$

SOLUTIONS

a. The denominator of 2 was multiplied by 4 to make a denominator of 8.

   Since the denominator was multiplied by 4, the numerator must also be multiplied by 4. The missing number is 4.

$$\frac{1}{2} = \frac{\square}{8}$$
$$\overset{\times\,4}{\frac{1}{2}} = \overset{\times\,4}{\frac{4}{8}}$$

b. The numerator of 3 was multiplied by 2 to make a numerator of 6.

   Since the numerator was multiplied by 2, you must multiply the denominator by 2.

   The missing number is 16.

$$\overset{\times\,2}{\frac{3}{8}} = \frac{6}{\square}$$
$$\frac{3}{8} = \underset{\times\,2}{\frac{6}{16}}$$

EXPLORE!　　　　　　　　　**CREATING EQUIVALENT FRACTIONS**

Step 1: Choose a fraction from the purple box. Create a model of this fraction using fraction tiles. Draw or trace the fraction on paper. Write the fraction below the drawing.

$$\frac{3}{6}$$

$\frac{4}{8}$	$\frac{3}{12}$	$\frac{6}{10}$	$\frac{2}{8}$	$\frac{6}{12}$	$\frac{9}{12}$
$\frac{4}{6}$	$\frac{2}{10}$	$\frac{4}{12}$	$\frac{2}{12}$	$\frac{10}{12}$	$\frac{3}{6}$
$\frac{5}{10}$	$\frac{2}{6}$	$\frac{4}{10}$	$\frac{6}{8}$	$\frac{8}{12}$	$\frac{8}{10}$

Step 2: Use the fraction tiles to make an equivalent fraction with a smaller denominator. Record this on your paper to show that the two fractions are equivalent.

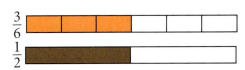

Step 3: Repeat **Steps 1-2** for four additional fractions from the purple box.

Step 4: Compare each of your fractions from the purple box to the equivalent fractions you formed. Were the original numerator and denominator multiplied or divided by a number to get the new fraction?

Step 5: Try to find equivalent fractions with smaller numerators and denominators for the fractions below without using a model.

$$\frac{10}{30} \qquad\qquad \frac{6}{12} \qquad\qquad \frac{15}{20}$$

EXAMPLE 2

Use division to find the missing number in the equivalent fractions.

a. $\dfrac{8}{16} = \dfrac{\square}{8}$ b. $\dfrac{12}{36} = \dfrac{1}{\square}$

SOLUTIONS

a. The denominator of 16 was divided
   by 2 to make a denominator of 8.

$$\dfrac{8}{16} = \dfrac{\square}{8}$$
$\div 2$

The denominator was divided by 2.
The numerator must also be divided by 2.

$$\dfrac{8}{16} = \dfrac{4}{8}$$
$\div 2$

The missing number is 4.

b. The numerator of 12 was divided
   by 12 to make a numerator of 1.

$$\dfrac{12}{36} = \dfrac{1}{\square}$$
$\div 12$

The numerator was divided by 12.
The denominator must also be divided by 12.

$$\dfrac{12}{36} = \dfrac{1}{3}$$
$\div 12$

The missing number is 3.

EXERCISES

1. Use models to determine if $\frac{1}{3}$ is equivalent to $\frac{2}{6}$.

   a. Copy the two rectangles below.

$\frac{1}{3}$ [ ]

$\frac{2}{6}$ [ ]

   b. Divide the first rectangle into three equal-sized sections. Divide the
   second rectangle into six equal-sized sections.
   c. Color in the number of sections given by the numerator of each fraction.
   d. Compare the two models. Are the fractions equivalent? Explain your answer.

Use models to show whether or not each pair of fractions is equivalent.

2. $\frac{1}{2}$ and $\frac{2}{4}$ **3.** $\frac{2}{5}$ and $\frac{3}{6}$ **4.** $\frac{1}{4}$ and $\frac{3}{8}$

5. Serj decided to use rectangular fraction tiles to show whether or not $\frac{5}{8}$ is equivalent to $\frac{3}{4}$. Look at Serj's model below. Are the fractions equivalent? How do you know?

6. Use rectangular tiles to show whether or not $\frac{4}{6}$ is equivalent to $\frac{2}{3}$.

Use multiplication to find the missing number for each equivalent fraction.

7. $\frac{1}{2} = \frac{\square}{8}$

8. $\frac{5}{6} = \frac{\square}{30}$

9. $\frac{3}{5} = \frac{18}{\square}$

10. $\frac{2}{3} = \frac{14}{\square}$

11. $\frac{1}{10} = \frac{\square}{30}$

12. $\frac{6}{7} = \frac{60}{\square}$

Use division to find the missing number for each equivalent fraction.

13. $\frac{10}{60} = \frac{1}{\square}$

14. $\frac{24}{64} = \frac{3}{\square}$

15. $\frac{8}{10} = \frac{\square}{5}$

16. $\frac{25}{35} = \frac{\square}{7}$

17. $\frac{12}{54} = \frac{2}{\square}$

18. $\frac{12}{14} = \frac{6}{\square}$

Write two fractions that are equivalent to each fraction.

19. $\frac{1}{3}$

20. $\frac{6}{12}$

21. $\frac{4}{10}$

22. $\frac{2}{8}$

23. $\frac{6}{9}$

24. $\frac{10}{25}$

25. Jedediah told his mom that $\frac{3}{9}$ of his chores were completed. His mother asked, "Are you telling me that $\frac{1}{3}$ of your chores are done?" How should he answer this question? Explain your reasoning.

26. Mykesha gave away 24 of the 30 sticks of gum in her pack.
   a. Write a fraction to represent the portion of gum she gave away.
   b. Write an equivalent fraction with a smaller numerator and denominator to represent the portion of gum she gave away.

REVIEW

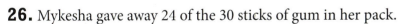

List the factors of each number. State whether the number is prime or composite.

27. 11

28. 24

29. 39

Find the GCF of each pair of numbers using any method.

30. 10 and 30

31. 20 and 24

32. 42 and 72

TIC-TAC-TOE ~ FRACTION GAME

Step 1: Draw a game board on a piece of poster board with a starting place and more than sixteen squares that lead to an ending place.

Step 2: Write at least 8 different fractions in simplest form, one on each square, not including the starting point. You may use the same fraction up to two times.

Step 3: Write 3 equivalent fractions on small cards for each simplified fraction you wrote on your game board. (There will be at least 24 cards.)

Step 4: Write directions for your game. *Examples:* Do players go backward if they draw a card that has a fraction that is equivalent to one behind them but none in front of them? How do they get to the ending place? Are there places on the game board where they are stuck until they can answer a specific question with fractions?

Step 5: Try the game with your family or friends to make sure it works.

Step 6: Bring your game to class to play.

TIC-TAC-TOE ~ EQUIVALENT FRACTIONS

Common fractions are fractions used in our world. For example, they are fractions used in cooking, sewing, woodworking, etc. Some examples of common fractions are:

$$\frac{1}{2} \qquad \frac{1}{4} \qquad \frac{3}{4} \qquad \frac{1}{3} \qquad \frac{2}{3} \qquad \frac{1}{8}$$

Write three equivalent fractions for each fraction above. Make a poster to display the sets of fractions. Use models to show each common fraction and its equivalent fractions.

SIMPLIFYING FRACTIONS

 Write fractions in simplest form.

A fraction is in simplest form when the numerator's and the denominator's only common factor is 1. Change a fraction into simplest form by repeatedly dividing by common factors until the only common factor between the numerator and denominator is 1.

EXPLORE!   **FRACTION HOMEWORK**

Five friends from different math classes worked on their fraction homework. The table shows the number of problems each student completed out of the total number assigned.

Name	Problems Completed	Problems Assigned
Marisol	12	20
Kevin	15	18
Oscar	30	40
Nancy	24	36
Julie	20	25

Step 1: Write a fraction to represent the portion of homework each student completed.

Step 2: Marisol wants to represent the portion of problems she has completed in simplest form. She needs to find a common factor of 12 and 20. What is one common factor of 12 and 20? Divide the numerator and denominator by this number to create an equivalent fraction.

Step 3: Look at the equivalent fraction you wrote in **Step 2**. Is there another common factor between the numerator and denominator of the fraction? If so, divide both parts of the fraction by this number to create another equivalent fraction. Continue doing this until the only common factor of the numerator and denominator is 1. This fraction is now in simplest form.

Step 4: Repeat this process for the other students. Write a fraction in simplest form to represent the portion of homework each student has completed.

EXAMPLE 1

Use common factors to write $\frac{24}{30}$ in simplest form.

SOLUTION

One common factor of 24 and 30 is 2.
Divide the numerator and denominator by 2.

$$\frac{24}{30} = \frac{12}{15}$$ (÷2, ÷2)

A common factor of 12 and 15 is 3.
Divide the numerator and denominator by 3.

$$\frac{12}{15} = \frac{4}{5}$$ (÷3, ÷3)

The only common factor between 4 and 5 is 1.
The fraction $\frac{4}{5}$ is in simplest form.

You can also use the greatest common factor to write a fraction in simplest form. Dividing the numerator and denominator by the greatest common factor will show the fraction in its simplest form.

WRITING FRACTIONS IN SIMPLEST FORM

Divide the numerator and denominator by common factors until the only common factor is 1.
OR
Divide the numerator and denominator by the greatest common factor (GCF).

EXAMPLE 2

Use the greatest common factor to write $\frac{10}{40}$ in simplest form.

SOLUTION

Find the factors of 10.   1, 2, 5, (10)
Find the factors of 40.   1, 2, 4, 5, 8, (10,) 20, 40

The GCF is 10.

Divide both the numerator and the
denominator by 10.

$$\frac{10}{40} = \frac{1}{4}$$ (÷10, ÷10)

The fraction $\frac{1}{4}$ is in simplest form.

EXAMPLE 3

Use the graph to determine the fraction of students who chose Track & Field as their favorite sport. Write the answer in simplest form.

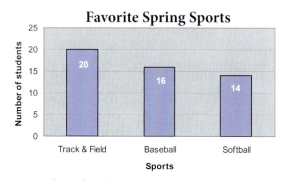

Favorite Spring Sports

SOLUTION

Write the fraction by taking the number of students who chose Track & Field out of the total number of students.

$$\frac{\text{number of students who chose Track \& Field}}{\text{total number of students}}$$

Find the total number of students surveyed.   20 + 16 + 14 = 50

Fifty students were surveyed. The denominator will be 50.

EXAMPLE 3
SOLUTION
(CONTINUED)

Twenty students said Track & Field was their favorite sport so 20 is the numerator.

$$\frac{20}{50}$$

Simplify the fraction. Use the common factor method or the GCF method.

Common Factors

$$\frac{20}{50} \xrightarrow{\div 5} = \frac{4}{10} \xrightarrow{\div 2} = \frac{2}{5}$$

The only common factor of 2 and 5 is 1. The fraction $\frac{2}{5}$ is in simplest form.

GCF

GCF = 10

$$\frac{20}{50} \xrightarrow{\div 10} = \frac{2}{5}$$

Two-fifths of the students said Track & Field is their favorite sport.

EXERCISES

Use common factors to write each fraction in simplest form. If the fraction is already in simplest form, write *simplest form*.

1. $\frac{9}{10}$

2. $\frac{15}{25}$

3. $\frac{16}{20}$

4. $\frac{48}{72}$

5. $\frac{32}{61}$

6. $\frac{30}{80}$

Use the GCF to write each fraction in simplest form. If it is already in simplest form, write *simplest form*.

7. $\frac{5}{15}$

8. $\frac{18}{30}$

9. $\frac{13}{18}$

10. $\frac{21}{28}$

11. $\frac{22}{77}$

12. $\frac{16}{28}$

Write each fraction in simplest form. Use any method.

13. $\frac{12}{20}$

14. $\frac{40}{60}$

15. $\frac{15}{45}$

16. $\frac{16}{24}$

17. $\frac{6}{16}$

18. $\frac{28}{42}$

Write each fraction in simplest form. Use any method.

19. $\frac{6}{15}$

20. $\frac{21}{84}$

21. $\frac{24}{27}$

Write each fraction in simplest form. State whether or not each pair of fractions is equivalent.

22. $\frac{10}{15}$ and $\frac{20}{30}$

23. $\frac{12}{24}$ and $\frac{16}{36}$

24. $\frac{14}{42}$ and $\frac{10}{36}$

25. Fractions can be simplified using two different methods. One method involves dividing the numerator and denominator by the GCF. The other method requires dividing the numerator and denominator by common factors until the only common factor that remains between the numerator and denominator is 1. Which method do you prefer to use? Why?

26. A snake measured $\frac{20}{25}$ meter long. Simplify this measurement.

27. Ochen weighed his club sandwich and found that it weighed $\frac{6}{16}$ pound. Simplify this measurement.

28. A student claims that $\frac{21}{41}$ is in simplest form. Do you agree? Explain.

Use the following graph. Write each fraction in simplest form.

29. What fraction of students chose to read *Charlie and the Chocolate Factory*?

30. What fraction of students chose to read *A Wrinkle in Time*?

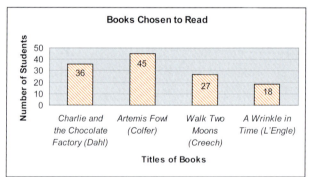

REVIEW

Write two fractions that are equivalent to each given fraction.

31. $\frac{1}{3}$

32. $\frac{9}{27}$

33. $\frac{40}{50}$

Use multiplication or division to find the missing number in each pair of equivalent fractions.

34. $\frac{1}{3} = \frac{4}{\square}$

35. $\frac{18}{27} = \frac{\square}{3}$

36. $\frac{22}{42} = \frac{11}{\square}$

LEAST COMMON MULTIPLE

LESSON 3.4

Find the least common multiple for a set of numbers.
Find the least common denominator for a set of fractions.

Dean and Kai volunteered together today at the recreation center. Dean volunteers every four days. Kai volunteers every five days. When will they be at the recreation center together again?

Dean's volunteer days ☐

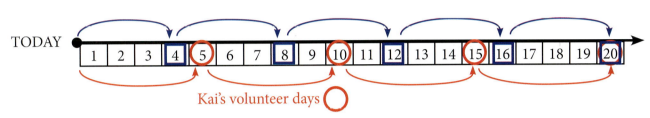

Kai's volunteer days ○

Look at the model above. The multiples of 4 are in blue squares and the multiples of 5 are circled in red. A **multiple** is the product of a given number and a whole number. The **least common multiple (LCM)** is the smallest multiple (not zero) that is common to two or more given numbers. By using the LCM, you can see that Dean and Kai will volunteer together in 20 days. That is the first time the ○ and ☐ are on the same number.

LEAST COMMON MULTIPLE (LCM)

List the multiples of the given numbers until you find the first multiple that is common to all given numbers.

OR

Use prime factorization of the given numbers to identify prime factors. Identify the common prime factors. Find the product of the common prime factors and any remaining factors.

EXAMPLE 1

List the first five non-zero multiples of 3 and 6. Find the LCM.

SOLUTION

Write the multiples of 3 and 6.

$3 \times 1 = 3$
$3 \times 2 = 6$
$3 \times 3 = 9$ **Multiples of 3**
$3 \times 4 = 12$
$3 \times 5 = 15$

$6 \times 1 = 6$
$6 \times 2 = 12$
$6 \times 3 = 18$ **Multiples of 6**
$6 \times 4 = 24$
$6 \times 5 = 30$

The first number to appear in both lists is 6.

List the multiples for each number.

Multiples of 3: 3, ⑥ 9, 12, 15…
Multiples of 6: ⑥ 12, 18, 24, 30…

The least common multiple (LCM) is 6.

EXAMPLE 2

Courtney and Kaysha are cousins. Courtney visits her grandparents every 30 days. Kaysha visits her grandparents every 20 days. They both visited their grandparents today. How long will it be before they are together at their grandparents' house on the same day?

SOLUTION

List the multiples of each number.

20: 20, 40, 60, 80, 100, 120, 140, 160, 180, ...

30: 30, 60, 90, 120, 150, 180, ...

Find the common multiples.
Common multiples of 20 and 30: 60, 120, 180

Identify the LCM.   LCM = 60

Courtney and Kaysha will be at their grandparents' house together in 60 days.

You can also use prime factorization to find the LCM of two numbers.

EXAMPLE 3

Sean and Deion both rented movies today. Sean rents a movie every 10 days. Deion rents a movie every 15 days. How many days until they rent movies on the same day again?

SOLUTION

Write the prime factorization for each number.

Identify the common prime factors (circled).

Find the LCM by finding the product of the common prime factors one time and any remaining prime factors.

LCM = $2 \times 3 \times 5 = 30$.

$$10 \qquad 15$$
$$2 \times 5 \qquad 3 \times 5$$
$$2 \times 3 \times 5 = 30$$

Sean and Deion will rent movies on the same day again in 30 days.

EXAMPLE 4

Ling, Kara and Sasha saw each other at the ice skating rink today. Ling goes to the rink every 2 days. Kara goes to the rink every 4 days. Sasha goes to the rink every 6 days. How many days will it be until they are together at the rink again?

SOLUTION

List multiples of each number.

2: 2, 4, 6, 8, 10, 12
4: 4, 8, 12, 16, 20, 24
6: 6, 12, 18, 24, 30, 36

Identify the LCM.   LCM = 12

Ling, Kara and Sasha will be at the rink together in 12 days.

The **least common denominator (LCD)** is the least common multiple (LCM) of two or more denominators.

EXAMPLE 5

Find the least common denominator of $\frac{3}{8}$ and $\frac{1}{2}$.

SOLUTION

Identify the denominators.   8 and 2

The LCM is the LCD.

List the multiples of both denominators.   8: 8, 16, 24, 32, 40
                                                                         2: 2, 4, 6, 8, 10, 12, 14, 16

The least common denominator is 8.

EXERCISES

List the first five non-zero multiples for each number.

1. 2   **2.** 6   **3.** 14

Use a list of multiples to find the LCM of each set of numbers.

4. 2 and 3   **5.** 4 and 10   **6.** 3 and 12

7. 6 and 10   **8.** 3, 5 and 6   **9.** 4, 8 and 12

Use prime factorization to find the LCM of each set of numbers.

10. 4 and 6   **11.** 10 and 14   **12.** 16 and 18

13. 14 and 21   **14.** 20 and 36   **15.** 5, 15 and 20

Find the LCD of each set of fractions.

16. $\frac{1}{4}, \frac{3}{5}$ **17.** $\frac{2}{5}, \frac{3}{7}$ **18.** $\frac{7}{8}, \frac{3}{14}$

19. $\frac{2}{7}, \frac{11}{14}$ **20.** $\frac{3}{9}, \frac{9}{15}$ **21.** $\frac{7}{8}, \frac{3}{12}$

22. Chad and Mike both delivered pizza for Pizza Ritza today. Chad works every 3 days. Mike works every 4 days. How many days until Chad and Mike work together again?

23. Tristan and Hayley get their digital photos printed regularly. They saw each other on Tuesday at the store. Tristan gets her photos printed every 2 days. Hayley gets her photos printed every 5 days. On which day will Tristan and Hayley be at the store on the same day again?

24. Diego, Jordan and Scott met each other at an open gym session today. Diego goes to open gym every 4 days. Jordan goes every 5 days. Scott goes every 6 days. How many days until all three boys go to open gym session on the same day?

25. On Thursday, Matt, Ryan and Alina saw each other at the library. Matt goes to the library to study every 6 days. Ryan goes every 12 days. Alina goes every 4 days. On which day of the week will they all be at the library at the same time?

REVIEW

Find the GCF for each pair of numbers.

26. 12 and 18

27. 15 and 45

28. 35 and 49

Use multiplication or division to find the missing number in each equivalent fraction expression.

29. $\frac{2}{8} = \frac{\square}{24}$

30. $\frac{7}{21} = \frac{\square}{3}$

31. $\frac{3}{5} = \frac{18}{\square}$

32. $\frac{24}{32} = \frac{3}{\square}$

33. $\frac{1}{6} = \frac{\square}{30}$

34. $\frac{27}{36} = \frac{3}{\square}$

TIC-TAC-TOE ~ EDIBLE FRACTIONS

Step 1: Start with a bag of multicolored candy (jelly beans, gum balls, gumdrops, etc.). Make sure there are at least four different colors of candy in the bag.

Step 2: Empty the candy bag. Count the total number of pieces of each color. Record how many there are of each color and the total of all colors combined.

Step 3: Write simplified fractions that represent each color as a portion of the candy.
Example: 125 pieces of red, yellow, purple or orange jelly beans

There are 45 red jelly beans. $\dfrac{\text{number of red jelly beans}}{\text{total number of candies}} = \dfrac{45}{125} = \dfrac{9}{25}$

Step 4: Put the fractions in order from least to greatest.

ORDERING AND COMPARING FRACTIONS

LESSON 3.5

Compare fractions with like and unlike denominators to find the smallest or largest fraction.

EXPLORE!

WHICH IS LARGER?

Darin, Yolanda and Ivan finished PE class and were thirsty. Darin drank $\frac{3}{5}$ liter of water. Yolanda drank $\frac{5}{8}$ liter of water. Ivan drank $\frac{7}{12}$ liter of water.

Step 1: Use fraction tiles to model each fraction in the situation above. Draw a picture of each model.

Step 2: Explain how you can tell which student drank the most water by looking at the models.

Step 3: List the students in order from who drank the least water to who drank the most water.

Step 4: Uma ate $\frac{1}{6}$ pound of carrots. Her brother ate $\frac{2}{12}$ pound of carrots. Use the fraction tiles to determine who ate the most carrots. Explain your answer.

Step 5: Diana's hair is $\frac{5}{6}$ foot long. Mikayla's hair is $\frac{2}{3}$ foot long. Larry's hair is $\frac{7}{8}$ foot long. List the people in order from shortest hair to longest hair.

Step 6: Explain how using fraction tiles can help with ordering and comparing fractions.

In a recent survey, three-fifths of people liked peanut butter and jam sandwiches. One-third of people liked grilled cheese. Which food is more popular?

JPD Incorporated surveyed its employees and found out that $\frac{1}{2}$ of the employees drank coffee. Tea drinkers made up $\frac{1}{6}$ of the employees. One-third of the employees preferred water. Which beverage was least popular?

In each of these situations, the fractions have different denominators. To compare fractions without using models, the denominators of the fractions need to be equal.

COMPARE AND ORDER FRACTIONS WITH UNLIKE DENOMINATORS

1. Find the least common denominator (LCD) for the fractions in the set.
2. Change each fraction to an equivalent fraction using the LCD.
3. Compare the numerators.

EXAMPLE 1

In a recent survey, three-fifths of the people liked peanut butter and jam. One-third liked grilled cheese sandwiches. Compare three-fifths and one-third to find which food was most popular.

SOLUTION

Convert words to numbers.

$\frac{3}{5}$ and $\frac{1}{3}$

The denominators are 3 and 5. List multiples of each.

3: 3, 6, 9, 12, ⑮
5: 5, 10, ⑮

LCM = 15

The LCM = 15, so the LCD = 15.

Make equivalent fractions with denominators of 15.

$\frac{3}{5} = \frac{9}{15}$ (×3)

$\frac{1}{3} = \frac{5}{15}$ (×5)

Compare the numerators. Nine is larger than 5.

$\frac{9}{15} > \frac{5}{15}$

> means "greater than"
< means "less than"

Substitute the original fractions in the comparison.

If $\frac{9}{15} > \frac{5}{15}$, then $\frac{3}{5} > \frac{1}{3}$.

The group that liked peanut butter and jam sandwiches was larger.

EXAMPLE 2

List the following fractions from least to greatest: $\frac{1}{2}, \frac{1}{6}, \frac{1}{3}$.

SOLUTION

List the fractions that need to be compared.

$\frac{1}{2}, \frac{1}{6}, \frac{1}{3}$

The denominators are 2, 6 and 3. List multiples of each.

2: 2, 4, ⑥, 8
3: 3, ⑥, 9
6: ⑥, 12, 18

The LCM = 6, so the LCD = 6.

Make equivalent fractions.

$\frac{1}{2} = \frac{3}{6}$ (×3) $\frac{1}{3} = \frac{2}{6}$ (×2)

Compare numerators and put them in order from least to greatest.

$\frac{1}{6} < \frac{2}{6} < \frac{3}{6}$

Substitute the original fractions for each simplified fraction to answer the question.

$\frac{1}{6}, \frac{1}{3}, \frac{1}{2}$

EXAMPLE 3

Find a fraction between $\frac{1}{6}$ and $\frac{1}{2}$. Write in simplest form.

SOLUTION

The denominators are 6 and 2.

$\frac{1}{6}, \frac{1}{2}$

List multiples of each denominator.

2: 2, 4, ⑥, 8
6: ⑥ 12, 18

The LCM = 6, so the LCD = 6.

Make equivalent fractions.

$\frac{1}{2} = \frac{3}{6}$ (×3)

Compare the fractions.

$\frac{1}{6} < \frac{3}{6}$

Write a fraction that goes between the two fractions.

$\frac{2}{6}$ would go between $\frac{1}{6}$ and $\frac{3}{6}$

Write the fraction in simplest form.

$\frac{2}{6} = \frac{1}{3}$

> Do not forget that all answers should be written in simplest form.

EXERCISES

Write the fraction for each drawing. Circle the largest fraction in each pair.

1.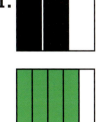

2.

Compare each pair of fractions. Replace each ● with <, > or = to make a true sentence.

3. $\frac{3}{5}$ ● $\frac{3}{4}$

4. $\frac{5}{8}$ ● $\frac{2}{3}$

5. $\frac{3}{12}$ ● $\frac{6}{24}$

6. $\frac{3}{10}$ ● $\frac{1}{4}$

7. $\frac{5}{12}$ ● $\frac{2}{5}$

8. $\frac{5}{7}$ ● $\frac{3}{5}$

9. $\frac{1}{4}$ ● $\frac{1}{3}$

10. $\frac{3}{5}$ ● $\frac{7}{12}$

11. $\frac{6}{11}$ ● $\frac{1}{2}$

Write each set of fractions in order from least to greatest.

12. $\frac{1}{2}, \frac{3}{4}, \frac{1}{4}$

13. $\frac{2}{5}, \frac{1}{3}, \frac{1}{5}$

14. $\frac{3}{8}, \frac{5}{8}, \frac{1}{2}$

15. $\frac{2}{5}, \frac{7}{10}, \frac{3}{10}$

16. $\frac{5}{6}, \frac{2}{6}, \frac{1}{4}$

17. $\frac{11}{15}, \frac{1}{3}, \frac{4}{5}$

Find a fraction between each pair of fractions. Write in simplest form.

18. $\frac{1}{3}, \frac{5}{8}$

19. $\frac{1}{3}, \frac{4}{5}$

20. $\frac{3}{8}, \frac{3}{5}$

21. $\frac{2}{9}, \frac{3}{7}$

22. $\frac{1}{5}, \frac{1}{3}$

23. $\frac{4}{9}, \frac{2}{3}$

24. Hani was making a cake that called for $\frac{5}{8}$ cup of cocoa and $\frac{2}{3}$ cup of butter. Would he need more cocoa or butter?

25. Worker honey bees are approximately $\frac{2}{3}$ inch long. Drone honey bees are approximately $\frac{1}{6}$ inch long. Which are longer?

26. Cody dumped out a box of gum balls. Two-ninths of the gum balls were pink. One-third of the gum balls were green. Four-ninths of the gum balls were white.
    a. Put the fractions in order from least to greatest.
    b. Cody had the most of which color of gum ball?

27. Kabira bought three candles that were the same size. After a few weeks she noticed she had burned the yellow candle $\frac{5}{12}$ of the way down. The red candle had burned $\frac{1}{3}$ of the way. The green candle had burned down $\frac{3}{8}$ of the way. Which candle had burned down the most?

REVIEW

Find the LCM for each pair of numbers.

28. 6 and 8

29. 12 and 15

30. 9 and 5

Use multiplication or division to find the missing number in each equivalent fraction expression.

31. $\frac{4}{7} = \frac{16}{\square}$

32. $\frac{15}{19} = \frac{\square}{38}$

33. $\frac{25}{60} = \frac{5}{\square}$

34. $\frac{35}{56} = \frac{\square}{8}$

Write each fraction in simplest form. If it is already in simplest form, write *simplest form*.

35. $\frac{3}{30}$

36. $\frac{12}{14}$

37. $\frac{26}{31}$

38. $\frac{42}{70}$

TIC-TAC-TOE ~ FAVORITES

Step 1: Create a survey question and four options by filling in the blanks:

What is your favorite _____?

Option 1:_____

Option 2:_____

Option 3:_____

Option 4:_____

Step 2: Survey at least fifty people and record their answers.

Step 3: Create a bar graph to show your results.

Step 4: Write the results for each option as simplified fractions.

Example: What is your favorite color?

Option 1: Red $\frac{20 \text{ people}}{50 \text{ people}} = \frac{2}{5}$

Step 5: Order the fractions from least to greatest.

Step 6: Write a paragraph summarizing the results of the survey.

Step 7: Create a poster using **Steps 1-6** to display your process.

Write improper fractions as mixed numbers.
Write mixed numbers as improper fractions.

A proper fraction has a numerator that is less than the denominator.

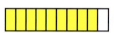

$\frac{9}{10}$ (nine-tenths) $\frac{2}{3}$ (two-thirds)

An improper fraction has a numerator that is equal to or greater than the denominator. Whole numbers can be written as improper fractions by writing the whole number over 1.

$\frac{11}{10}$ (eleven-tenths) $\frac{6}{3}$ or $\frac{2}{1}$ or 2

A mixed number is the sum of a whole number and a fraction.

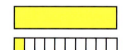

$1\frac{1}{10}$ (one and one-tenth) $1\frac{1}{3}$ (one and one-third)

A number line can be used to show improper fractions and their equivalent mixed numbers.

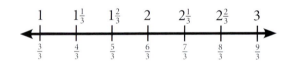

REWRITING IMPROPER FRACTIONS AS MIXED NUMBERS

1. Divide the numerator by the denominator. The quotient is the whole number in the mixed number.
2. Write the remainder as the numerator over the original denominator. This is the fraction in the mixed number.

EXAMPLE 1   **Change the improper fraction $\frac{14}{4}$ to a mixed number.**

SOLUTION

Divide the numerator by the denominator.
The quotient is the whole number.
The remainder goes over the original denominator to make the fraction.

$$\frac{14}{4} = 4\overline{\smash{)}14} = 3\frac{2}{4}$$
$$\underline{-12}$$
Remainder: 2

Simplify.

$$3\frac{2}{4} = 3\frac{1}{2}$$

REWRITING MIXED NUMBERS AS IMPROPER FRACTION

1. Multiply the whole number by the denominator.
2. Add the numerator to the product.
3. Write this number as the numerator and keep the original denominator as the denominator of the improper fraction.

EXAMPLE 2

Change the mixed number $2\frac{3}{4}$ to an improper fraction.

SOLUTION

Multiply the whole number by the denominator.
Add the numerator to the product.

$$2\frac{3}{4} = \frac{4 \times 2 + 3}{4} = \frac{11}{4}$$

This number becomes the numerator.
The denominator stays the same.

$$2\frac{3}{4} = \frac{11}{4}$$

EXPLORE!   **CHOCOLATE CHIP COOKIES**

Lynn got out a recipe for cookies. She was confused when she looked at the amounts. The entire recipe was written with fractions. Most of them were improper fractions. Help her simplify the recipe by creating a new recipe card.

Chocolate Chip Cookies

$\frac{9}{8}$ cups butter $\frac{19}{8}$ cups flour

$\frac{6}{8}$ cup sugar $\frac{5}{4}$ tsp. baking soda

$\frac{6}{8}$ cup brown sugar $\frac{9}{8}$ tsp. salt

$\frac{8}{4}$ eggs $\frac{18}{8}$ cups chocolate chips

$\frac{10}{8}$ tsp. vanilla

Step 1: Look at the amount of butter on the recipe card. Draw a model of this fraction on paper. Write it as a mixed number on a blank recipe card.

Step 2: Draw a model of the fraction that represents the amount of sugar in the cookies on your paper. Write it as a simplified fraction on the new recipe card.

Step 3: Continue to use the drawings to figure out the mixed number, whole number or simplified fraction for each ingredient. Draw models for each of these fractions on your paper and record the fraction, whole number or mixed number on your recipe card.

Step 4: Lynn wanted to confuse her brother by changing a recipe for brownies into improper, non-simplified fractions. Create a recipe card from the brownie recipe below to confuse him.

Brownies			
$\frac{1}{2}$ cup butter	$\frac{1}{4}$ cup sugar	2 eggs	$1\frac{1}{2}$ tsp. vanilla
$\frac{1}{3}$ cup unsweetened cocoa	$1\frac{1}{8}$ cups flour	$\frac{1}{4}$ tsp. salt	$\frac{3}{8}$ tsp. baking powder

EXERCISES

Write a mixed number and improper fraction that represents the shaded amount for each drawing.

1.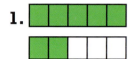

2.

3.

Write each improper fraction as a mixed number in simplest form.

4. $\frac{9}{2}$

5. $\frac{7}{3}$

6. $\frac{24}{5}$

7. $\frac{20}{6}$

8. $\frac{22}{9}$

9. $\frac{32}{7}$

10. $\frac{5}{2}$

11. $\frac{40}{12}$

12. $\frac{44}{14}$

13. The world's largest functioning mobile phone is the Maxi Mobile, according to the Guinness Book of World Records. It measures $\frac{336}{50}$ feet tall. Write this improper fraction as a mixed number in simplest form.

14. The average weight of a newborn baby is $\frac{117}{16}$ pounds. Write this improper fraction as a mixed number.

Write each mixed number as an improper fraction.

15. $4\frac{2}{3}$

16. $1\frac{5}{6}$

17. $2\frac{4}{7}$

18. $3\frac{3}{5}$

19. $5\frac{1}{8}$

20. $4\frac{3}{10}$

21. $6\frac{3}{9}$

22. $8\frac{1}{2}$

23. $3\frac{9}{15}$

24. According to the Guinness Book of World Records, the world's longest skateboard measured $30\frac{1}{12}$ feet long. Write this mixed number as an improper fraction.

Write each set of improper fractions and mixed numbers in order from least to greatest.

25. $\frac{7}{2}, 3\frac{1}{4}, \frac{9}{2}$

26. $\frac{8}{3}, 2\frac{1}{6}, \frac{10}{3}$

27. $\frac{7}{4}, 1\frac{1}{8}, \frac{3}{2}$

28. Your class is planning an ice cream party. There are six ice cream bars in each box. There are twenty-three students in your class. Each student will eat one ice cream bar. Write a mixed number representing how many boxes of ice cream bars your class will eat.

29. Which is longer, $10\frac{5}{8}$ inches or $\frac{40}{4}$ inches?

30. It rained $1\frac{7}{8}$ inches on Monday. On Tuesday, it rained $\frac{7}{4}$ of an inch. On which day did it rain more?

REVIEW

Find the value of each expression.

31. 3.39×5.22

32. $92.5 \div 8$

33. 15.9×10.2

Compare each pair of fractions. Replace each ⬤ with <, > or = to make a true sentence.

34. $\frac{5}{7}$ ⬤ $\frac{3}{4}$

35. $\frac{2}{3}$ ⬤ $\frac{2}{5}$

36. $\frac{4}{10}$ ⬤ $\frac{6}{15}$

Find the GCF of each pair of numbers.

37. 36 and 18

38. 8 and 16

39. 10 and 55

TIC-TAC-TOE ~ RECIPE MIX-UP

Choose six of your favorite recipes. Each recipe must have a minimum of five ingredients.

Change all of the numbers and fractions in the recipe by using an equivalent fraction for each fraction given. Write all whole numbers and mixed numbers as improper fractions.

Examples: "2 cups flour" can be written as "$\frac{10}{5}$ cups flour"

"$\frac{3}{4}$ cup brown sugar" can be written as "$\frac{9}{12}$ cup brown sugar"

"$1\frac{1}{2}$ tsp. salt" can be written as "$\frac{3}{2}$ tsp. salt"

Create a "My Favorite Recipes" cookbook using the recipes with mixed-up fractions. Include directions with each recipe.

MEASURING IN INCHES

LESSON 3.7

 Use a customary ruler to measure inches and fractions of an inch.

A jumbo paper clip is about $\frac{7}{8}$ inch long.

A pen is about $5\frac{1}{2}$ inches long.

A pencil sharpener is about $1\frac{1}{4}$ inches long.

In many careers it is important to know how to measure accurately using a ruler. A carpet installer needs to know the length and width of a room to order the correct amount of carpet. A builder needs to measure wooden beams so they are the same length. A plumber needs to measure the length of pipe accurately. A ruler measures inches using different sized denominators like the examples above. It is important to know what each line on a ruler represents.

A foot-long ruler has 12 inches. Each inch is separated into sixteenths ($\frac{1}{16}$) on the ruler using different sized lines, called tick marks. There are 16 equally divided spaces between every inch on a ruler.

EXPLORE!   **USING A CUSTOMARY RULER**

Step 1: Copy and complete the table below by writing each fraction in simplest form.

Tick Mark	1	2	3	4	5	6	7	8	9	10	11	12	13	14	15	16
Fraction of Inch	$\frac{1}{16}$	$\frac{2}{16}$	$\frac{3}{16}$	$\frac{4}{16}$	$\frac{5}{16}$											
Simplest Form																

Step 2: How many fractions in the table above simplify to a whole number? What do you notice about the tick mark(s) on the ruler that correspond to the measurement(s) that are whole numbers?

Step 3: After simplifying the fractions, how many fractions have a 2 in the denominator? What do you notice about the tick marks on the ruler that correspond to the measurements with a 2 in the denominator?

Step 4: How many of the simplified fractions have a 4 in the denominator? What do you notice about the tick marks on the ruler that correspond to the measurements with a 4 in the denominator?

Step 5: How many fractions have an 8 in the denominator? What do you notice about the tick marks on the ruler that correspond to the measurements with an 8 in the denominator?

Step 6: After simplifying the fractions, how many fractions have a 16 in the denominator? What do you notice about the tick marks on the ruler that correspond to the measurements with a 16 in the denominator?

It is important to measure accurately when measuring the length of a wall in a room or the height of a person at the doctor's office. To measure length as accurately as possible, measure to the nearest sixteenth of an inch.

> ### MEASURING WITH A CUSTOMARY RULER
>
> 1. Locate the tick mark that most closely matches the length of the object.
> 2. Record the whole number of inches.
> 3. Record the fractional part of an inch that the tick mark represents.
> 4. Write the whole number of inches with the fraction part of an inch in simplest form.

EXAMPLE 1

Measure each line using inches on a ruler. Measure to the nearest sixteenth of an inch.

a. ————————————————

b. ——————————————

SOLUTIONS

For both measurements, line up the "0" inch mark of the ruler with the left edge of the line drawn. Identify which tick mark on the ruler corresponds to the right end of the line.

a.

The line corresponds to 3 whole inches plus 2 additional tick marks. $3 + \frac{2}{16} = 3\frac{2}{16}$

Write the fraction in simplest form. $3\frac{2}{16} = 3\frac{1}{8}$ inches

b.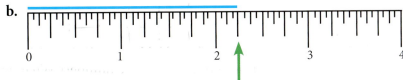

The line corresponds to 2 whole inches plus 4 additional tick marks.
$$2 + \frac{4}{16} = 2\frac{4}{16}$$

Write the fraction in simplest form. $2\frac{4}{16} = 2\frac{1}{4}$ inches.

Sometimes measurements will be rounded to the nearest quarter $\left(\frac{1}{4}\right)$ inch or half $\left(\frac{1}{2}\right)$ inch when exact answers are not required.

EXAMPLE 2

Measure the line to the nearest quarter inch.

_____

SOLUTION

Measuring to the nearest quarter inch means the answer could include
$\frac{1}{4}$ inch, $\frac{2}{4}$ inch, $\frac{3}{4}$ inch, or $\frac{4}{4}$ inch

Notice: $\frac{2}{4} = \frac{1}{2}$
and $\frac{4}{4} = 1$

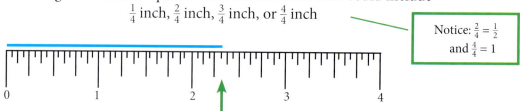

Use a ruler to see that the line is 2 inches and then ends closer to the $\frac{1}{4}$ tick mark than to the $\frac{1}{2}$ tick mark.

The line is about $2\frac{1}{4}$ inches long.

EXERCISES

1. The tick marks divide each inch into how many sections?

2. How many inches are in 1 foot?

Round to the nearest half inch.

3. $3\frac{1}{16}$ in

4. $4\frac{5}{8}$ in

5. $9\frac{7}{8}$ in

Round to the nearest quarter inch.

6. $2\frac{3}{16}$ in

7. $10\frac{15}{16}$ in

8. $4\frac{7}{16}$ in

Measure the length of each line to the nearest sixteenth of an inch.

9. _____

10. _____

11. _____

12. _____

Measure the length of each line to the nearest quarter of an inch.

13. _____

14. _____

15. _____

16. Kim and Marci measured the width of the table they sat at each day at school. The width of the table was $25\frac{9}{16}$ inches. The math teacher asked the students to measure the width of the table to the nearest quarter inch. Kim answered $25\frac{3}{4}$ inches. Marci insisted the answer was $25\frac{1}{2}$ inches.
  a. Which answer is closest to $25\frac{9}{16}$ inches?
  b. Kim was sure the answer could not end in $\frac{1}{2}$ if they were supposed to round to the nearest quarter inch. Explain why the answer can or cannot end in $\frac{1}{2}$ when rounding to the nearest quarter inch.

Draw a line that has the given length.

17. $\frac{3}{8}$ inch

18. $5\frac{1}{4}$ inches

19. $2\frac{1}{2}$ inches

20. $3\frac{9}{16}$ inches

21. 4 inches

Determine which measurement is longer.

22. $\frac{1}{8}$ inch or $\frac{1}{4}$ inch

23. $\frac{1}{2}$ inch or $\frac{7}{16}$ inch

24. $\frac{5}{8}$ inch or $\frac{3}{4}$ inch

25. Jon had three wires. The red wire was $\frac{1}{4}$ inch long. The blue wire was $\frac{3}{8}$ inch long. The green wire was $\frac{3}{12}$ inch long. Which two wires are equal in length?

26. Martina has a piece of purple yarn that is $3\frac{1}{2}$ inches long. She has a piece of yellow yarn that is $3\frac{7}{12}$ inches long. Which piece of yarn is longer?

27. Antonio measured his hand from his wrist to the tip of his longest finger. It was $7\frac{3}{4}$ inches long. His brother's hand measured $7\frac{7}{8}$ inches long.
  a. Whose hand length is shortest?
  b. Measure the length of your hand. Is it longer or shorter than Antonio's hand?

REVIEW

Write each set of fractions in order from least to greatest.

28. $\frac{1}{2}, \frac{3}{16}, \frac{7}{16}$

29. $\frac{4}{5}, \frac{7}{10}, \frac{2}{5}$

30. $\frac{7}{9}, \frac{11}{18}, \frac{20}{27}$

Write each improper fraction as a mixed number.

31. $\frac{17}{9}$

32. $\frac{16}{5}$

33. $\frac{9}{4}$

TIC-TAC-TOE ~ COINS AND FRACTIONS

A coin can be written as a fraction of a dollar.

Example: A fifty cent piece.

There are 50 cents in one fifty cent piece and one hundred cents in a dollar.

One fifty cent piece is $\frac{50 \text{ cents}}{100 \text{ cents}} = \frac{50}{100} = \frac{1}{2}$ of a dollar.

This makes sense because a fifty cent piece is one half of a dollar.

Step 1: Copy and complete the first five rows on the table below to show the value of each coin as a fraction of a dollar. Add ten blank rows to the table.

Coin	Value as a fraction
Fifty cent piece	$\frac{1}{2}$ of a dollar
Quarter	
Dime	
Nickel	
Penny	

Step 2: Put nine quarters, nine dimes, nine nickels, and nine pennies into a bag. Reach into the bag and pull out three coins. Record the coins in the first empty row on your table. Figure out the combined value as a fraction. Write the fraction, in simplest form, in the table.

Example: 1 quarter, 2 pennies is $\frac{27}{100}$ of a dollar.

Step 3: Put the coins back into the bag. Reach in and pull out three-eight coins. Record these coins in another row on your table. Figure out the combined value as in **Step 2.**

Step 4: Repeat **Step 3** until you have completed ten more rows on the table.

Step 5: Write a quiz with at least ten questions about fractions of a dollar. Make at least half of your questions challenging.

Example: "Which three coins can be combined to make $\frac{40}{100}$ of a dollar?"

This question could be more challenging by simplifying the fraction in the question. Change the question to, "Which three coins could be combined to make $\frac{2}{5}$ of a dollar?"

Step 6: Make an answer key for your quiz.

Vocabulary

composite number least common denominator prime factorization

equivalent fractions   least common multiple   prime number

greatest common factor mixed numbers   proper fraction

improper fraction   multiple   simplest form

Find the greatest common factor (GCF) of a set of numbers.
Write equivalent fractions.
Write fractions in simplest form.
Find the least common multiple for a set of numbers.
Find the least common denominator for a set of fractions.
Compare fractions with like and unlike denominators to find the smallest or largest fraction.
Write improper fractions as mixed numbers.
Write mixed numbers as improper fractions.
Use a customary ruler to measure inches and fractions of an inch.

Lesson 3.1 ~ Greatest Common Factor

• •

List the factors of each number. State whether each number is prime or composite.

1. 5   **2.** 9   **3.** 21

Find the greatest common factor (GCF) for each pair of numbers.

4. 15 and 20   **5.** 14 and 21   **6.** 12 and 30

7. 18 and 27   **8.** 10 and 25   **9.** 40 and 56

Lesson 3.2 ~ Equivalent Fractions

• •

Use models to show whether or not each pair of fractions is equivalent.

10. $\frac{1}{4}$ and $\frac{2}{8}$ **11.** $\frac{3}{8}$ and $\frac{5}{6}$

Use multiplication or division to find the missing number in each equivalent fraction expression.

12. $\frac{5}{9} = \frac{\square}{45}$ **13.** $\frac{1}{2} = \frac{4}{\square}$ **14.** $\frac{12}{36} = \frac{\square}{6}$

15. $\frac{3}{7} = \frac{21}{\square}$ **16.** $\frac{14}{56} = \frac{1}{\square}$ **17.** $\frac{18}{20} = \frac{\square}{10}$

Write two fractions that are equivalent to each fraction.

18. $\frac{3}{9}$

19. $\frac{6}{10}$

20. $\frac{1}{2}$

Lesson 3.3 ~ Simplifying Fractions

Write each fraction in simplest form. If it is already in simplest form, write *simplest form*.

21. $\frac{9}{12}$

22. $\frac{11}{15}$

23. $\frac{12}{15}$

24. $\frac{20}{30}$

25. $\frac{24}{81}$

26. $\frac{20}{37}$

27. Peter's dirt bike weighs $\frac{22}{200}$ ton. Simplify this measurement.

Write each fraction in simplest form. Tell whether or not the pair of fractions is equivalent.

28. $\frac{20}{35}$ and $\frac{8}{14}$

29. $\frac{12}{30}$ and $\frac{24}{50}$

Lesson 3.4 ~ Least Common Multiple

List the first five nonzero multiples for each number.

30. 3

31. 15

32. 10

Find the least common multiple (LCM) of each set of numbers.

33. 8 and 20

34. 9 and 12

35. 32 and 24

Find the least common denominator (LCD) of each set of fractions.

36. $\frac{5}{25}$, $\frac{2}{10}$

37. $\frac{3}{8}$, $\frac{5}{12}$

38. $\frac{4}{11}$, $\frac{2}{4}$

39. The Piper and Roy families like to hike. Both families went hiking today. Mr. Piper told Mr. Roy that the Piper family goes hiking every 15 days. Mr. Roy said his family goes hiking every 10 days. How many days until both families go hiking again on the same day?

Lesson 3.5 ~ Ordering and Comparing Fractions

Replace each ● with <, > or = to make a true sentence.

40. $\frac{5}{6}$ ● $\frac{9}{10}$

41. $\frac{4}{15}$ ● $\frac{1}{6}$

42. $\frac{3}{4}$ ● $\frac{12}{18}$

Write each set of fractions in order from least to greatest.

43. $\frac{2}{5}, \frac{3}{10}, \frac{3}{5}$

44. $\frac{1}{4}, \frac{3}{4}, \frac{2}{3}$

45. $\frac{4}{5}, \frac{3}{7}, \frac{2}{5}$

46. Levi had a candy bar. He ate $\frac{3}{8}$ of it. His sister ate $\frac{5}{16}$ of it. Who ate less?

47. Hoi read $\frac{7}{15}$ of her book on Monday. She read $\frac{4}{9}$ of her book on Tuesday. On which day did she read more of her book?

Lesson 3.6 ~ Mixed Numbers and Improper Fractions

Write each improper fraction as a mixed number.

48. $\frac{7}{2}$

49. $\frac{23}{5}$

50. $\frac{34}{11}$

Write each mixed number as an improper fraction.

51. $3\frac{5}{6}$

52. $2\frac{7}{11}$

53. $7\frac{5}{12}$

Write each set of improper fractions and mixed numbers in order from least to greatest.

54. $2\frac{1}{10}, \frac{11}{10}, 2\frac{3}{5}$

55. $1\frac{1}{3}, \frac{11}{6}, \frac{5}{3}$

56. $3\frac{1}{8}, \frac{17}{8}, \frac{7}{2}$

Lesson 3.7 ~ Measuring In Inches

Measure the length of each line to the nearest sixteenth of an inch.

57. _____

58. _____

59. _____

Measure the length of each line to the nearest quarter of an inch.

60. _____

61. _____

62. _____

Draw a line that has the given length.

63. $5\frac{1}{8}$ inches

64. $3\frac{1}{2}$ inches

65. $\frac{7}{16}$ inch

66. $2\frac{3}{4}$ inches

TIC-TAC-TOE ~ CONCAVE AND CONVEX

A concave figure is a closed figure that has an indentation.

A convex figure is a closed figure with no indentation.

Step 1: Draw seven different line segments with the following lengths onto a piece of paper. Leave room to cut each of them out.

    a. $2\frac{7}{8}$ inches   **b.** $1\frac{1}{4}$ inches   **c.** $\frac{15}{16}$ inch   **d.** $3\frac{1}{2}$ inches

    e. $1\frac{13}{16}$ inches   **f.** $2\frac{5}{8}$ inches   **g.** $4\frac{3}{16}$ inches

Step 2: Cut each line segment out carefully.

Step 3: Move the seven line segments to create a concave figure. Sketch this figure onto a piece of paper. Label the lengths of each side.

Step 4: Repeat **Step 3** until you have three or more concave figures.

Step 5: Move the seven line segments to create a convex figure. Sketch this figure onto a piece of paper. Label the lengths of each side.

Step 6: Repeat **Step 5** until you have three or more convex figures.

CAREER FOCUS

DIXIE METALSMITH

I am a metalsmith. My work has taken me in three directions. Currently I am teamed up with a wood sculptor and we design sculptures that combine wood and metal which we sell in art galleries. I also create one-of-a-kind jewelry. Finally, I teach metalworking to children ages nine to thirteen at a community center and in a summer arts program.

My work requires the use of many mathematical skills. Measuring accurately is critical in creating jewelry and sculpture. For small things I use the metric system, measuring in millimeters. For our sculptures, we use inches because we are working on a large scale. I weigh things using ounces and pennyweights (a common measure for weighing small amounts of silver, gold and other metals). I also use many formulas to figure circumferences of rings, bracelets, stones and geometric shapes (cones, cylinders and triangles) used in sculptures and jewelry. Mixing chemicals for cleaning and coloring the metal also require special formulas. I find metalworking uses most of the math skills I have learned over the years.

Jewelers and metalsmiths come from many different backgrounds. I have a degree in Art Education and also served an apprenticeship in a small jewelry/sculpture foundry. After that, I went to work in a few other art foundries, gaining more skills. Others may choose to get training through a technical school or distance learning class.

Salaries can vary in the jewelry field, but usually start at $8.00 - $15.00 an hour. With two years of experience in Portland, a jeweler can make $16.00 - $25.00 an hour, depending on expertise. There are jobs in small business and larger manufacturing situations. In the Portland area, the median salary for a job in this field is $34,000 per year. About forty percent of all jewelers are self-employed, so there are many opportunities to earn different amounts.

I love this profession because of all the skills it requires from the design to the completed piece. The business aspects of figuring cost, overhead, and sales is also challenging. Teaching provides an opportunity to pass on my knowledge and give others tools to express their ideas. My work with another artist gives me new direction and inspiration for growing in my profession.

CORE FOCUS ON DECIMALS & FRACTIONS
BLOCK 4 ~ ADDING AND SUBTRACTING FRACTIONS

LESSON 4.1   ESTIMATING SUMS AND DIFFERENCES ------------------------------------ 108

LESSON 4.2   ADDING AND SUBTRACTING FRACTIONS ------------------------------------ 112
            EXPLORE! PIZZA PARTY!

LESSON 4.3   ADDING AND SUBTRACTING MIXED NUMBERS --------------------------- 117
            EXPLORE! MIXING PAINT

LESSON 4.4   ADDING AND SUBTRACTING BY RENAMING --------------------------- 122

LESSON 4.5   PERIMETER WITH FRACTIONS --- 126

REVIEW   BLOCK 4 ~ ADDING AND SUBTRACTING FRACTIONS --------------------- 130

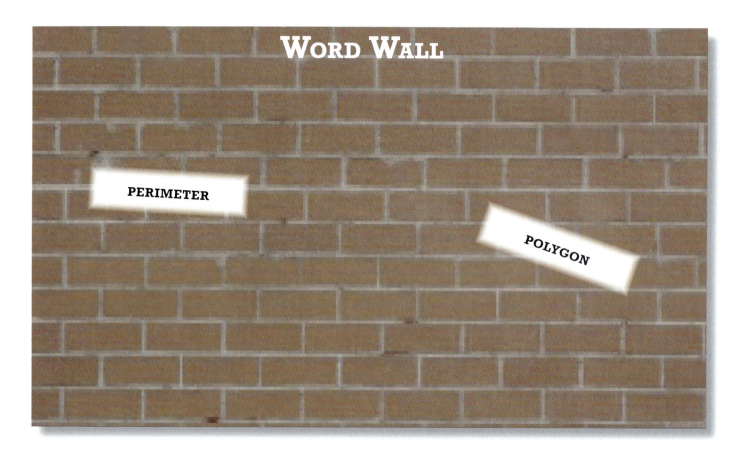

BLOCK 4 ~ ADDING AND SUBTRACTING FRACTIONS

TIC-TAC-TOE

LIGHTS, CAMERA, ACTION

Create a skit where characters use addition or subtraction of fractions to solve their problem.

See page 120 for details.

MEASURE THE PERIMETER

Measure the length and width of several different frames at home or at a store. Find each perimeter.

See page 129 for details.

CREATE THE PROBLEM

Write problems using pairs of fractions that add up to a given sum.

See page 125 for details.

MODELS

Create a brochure to teach how to add and subtract fractions using models.

See page 116 for details.

FIND THE SUM

Find the sum of problems of three or more fractions with different denominators.

$$\frac{1}{2} + \frac{1}{3} + \frac{1}{4}$$

See page 132 for details.

LANDSCAPING

Find the number of edging blocks you would need around each flower bed in a backyard.

See page 111 for details.

TRIATHLON

Find different distances using Oregon triathlon races.

See page 121 for details.

INTERVIEW

Interview someone who uses fractions. Report on how that person uses fractions.

See page 111 for details.

YOU ARE THE AUTHOR

Investigate children's books on fractions. Write your own children's book.

See page 120 for details.

ESTIMATING SUMS AND DIFFERENCES

LESSON 4.1

Estimate sums and differences of expressions with fractions and mixed numbers.

A teacher needed two boxes of pencils. Hayden said he had $\frac{1}{6}$ of a box at home. Trevor thought he had $\frac{2}{5}$ of a box. Kelsey knew she had $\frac{4}{7}$ of a box at home. Asha thought she had $\frac{5}{6}$ of a box. The students were not sure if they had enough pencils to help the teacher reach two boxes.

Estimating could help these students get a good idea of whether or not they had enough pencils.

ESTIMATING SUMS OR DIFFERENCES USING FRACTIONS

1. Round to 0, $\frac{1}{2}$ or 1, whichever is closest.
   ◆ If the numerator is very small compared to the denominator, estimate as 0.
   ◆ If the numerator is about half of the denominator, estimate as $\frac{1}{2}$.
   ◆ If the numerator is nearly as big as the denominator, estimate as 1.
2. Add or subtract.

EXAMPLE 1

Hayden had about $\frac{1}{6}$ of a box of pencils. Trevor had about $\frac{2}{5}$ of a box. Kelsey had about $\frac{4}{7}$ of a box and Asha had about $\frac{5}{6}$ of a box. About how many boxes of pencils do they have altogether?

SOLUTION

Compare each numerator to its denominator to determine the estimated value.

Hayden	Trevor	Kelsey	Asha
$\frac{1}{6} \rightarrow 0$	$\frac{2}{5} \rightarrow \frac{1}{2}$	$\frac{4}{7} \rightarrow \frac{1}{2}$	$\frac{5}{6} \rightarrow 1$
The numerator is very small compared to the denominator.	The numerators are about half of their denominators.		The numerator is nearly as big as the denominator.

Add the estimated amounts.
$$0 + \frac{1}{2} + \frac{1}{2} + 1 = 2$$

They have about two boxes of pencils among the four of them.
$$\frac{1}{6} + \frac{2}{5} + \frac{4}{7} + \frac{5}{6} \approx 2$$

The $\approx$ sign means "about" or "approximately."

EXAMPLE 2

Asha found that, instead of $\frac{5}{6}$ of a box of pencils, she had $\frac{4}{9}$ of a box. About how much less does she have than she originally thought?

SOLUTION

Write the problem.

$$\frac{5}{6} - \frac{4}{9} \approx$$

Round to 0, $\frac{1}{2}$ or 1.

$$1 - \frac{1}{2}$$

Subtract.

$$1 - \frac{1}{2} = \frac{1}{2} \quad \text{so} \quad \frac{5}{6} - \frac{4}{9} \approx \frac{1}{2}$$

Asha has about half of a box less than she thought she had.

ESTIMATING SUMS OR DIFFERENCES USING MIXED NUMBERS

1. Round to the nearest whole number.
2. Add or subtract.

EXAMPLE 3

Estimate the value of $2\frac{1}{3} + 5\frac{6}{7}$.

SOLUTION

Write the problem.

$$2\frac{1}{3} + 5\frac{6}{7}$$

Round to the nearest whole number.

$$2 + 6$$

Add.

$$2 + 6 = 8$$

$$2\frac{1}{3} + 5\frac{6}{7} \approx 8$$

EXAMPLE 4

According to the Guinness Book of World Records, the world's tallest man, Robert Pershing Wadlow, was $8\frac{11}{12}$ feet tall. The world's shortest woman, Zhu Haizhen, was $2\frac{7}{12}$ feet tall. Approximately how much taller was the tallest man than the shortest woman?

SOLUTION

Write the problem.

$$8\frac{11}{12} - 2\frac{7}{12} \approx$$

Round to the nearest whole number.

$$9 - 3$$

Subtract.

$$9 - 3 = 6$$

$$8\frac{11}{12} - 2\frac{7}{12} \approx 6$$

The tallest man was about six feet taller than the shortest woman.

EXERCISES

Estimate each sum or difference.

1. $\frac{2}{5} + \frac{2}{11}$

2. $\frac{3}{5} + \frac{4}{9}$

3. $\frac{9}{10} - \frac{7}{8}$

4. $\frac{8}{15} + \frac{6}{7}$

5. $\frac{32}{37} + \frac{21}{25}$

6. $\frac{7}{9} - \frac{6}{13}$

7. $\frac{19}{21} - \frac{1}{9}$

8. $\frac{8}{15} - \frac{10}{21}$

9. $\frac{9}{19} + \frac{2}{15}$

10. Melanie exercised for $\frac{7}{8}$ hour on Saturday. On Sunday she exercised for $\frac{3}{5}$ hour. About how many hours did Melanie exercise in the two days?

11. Sapphire and Rebekah each brought a pie to a family party. Sapphire cut her pie into 12 pieces. After the party, $\frac{1}{12}$ of the pie was left. Rebekah cut her pie into 8 pieces. She had $\frac{5}{8}$ of the pie left after the party. About how much more of Rebekah's pie was left than Sapphire's pie?

12. Nigel told his mom he completed $\frac{1}{10}$ of the family's chores. His brother, Nick, completed $\frac{4}{9}$ of the chores. Estimate what fraction of the chores were completed altogether.

Estimate each sum or difference.

13. $2\frac{2}{7} + 5\frac{8}{9}$

14. $2\frac{1}{5} + 3\frac{4}{7}$

15. $6\frac{8}{13} + 8\frac{3}{4}$

16. $9\frac{1}{2} + 1\frac{1}{3}$

17. $11\frac{2}{3} + 12\frac{7}{8}$

18. $6\frac{13}{14} - 4\frac{1}{8}$

19. $4\frac{8}{9} - 1\frac{3}{5}$

20. $5\frac{2}{5} - 2\frac{6}{11}$

21. $9\frac{2}{9} - 3\frac{7}{8}$

22. After soccer practice, Jaafan drank $1\frac{7}{8}$ cups of water. His friend, Brock, drank $1\frac{1}{7}$ cups of water. Approximately how much more water did Jaafan drink than Brock?

23. Trey's family drove from Redmond to Madras. The map showed that they had traveled $26\frac{3}{20}$ miles. They drove on to Warm Springs which was another $14\frac{4}{5}$ miles. Estimate the total number of miles they traveled from Redmond to Warm Springs.

24. Lacey made bread. The recipe called for $2\frac{3}{4}$ cups of flour at the beginning. She added an additional $3\frac{1}{6}$ cups of flour after mixing the first ingredients. Estimate how many total cups of flour Lacey needed to make the bread.

25. When the Jacobsens designed their house, one room was $16\frac{1}{8}$ feet wide. They changed the design so the width of the room was $1\frac{3}{4}$ feet shorter than originally planned. Approximately how wide would the new room be with the revised plan?

Write each mixed number as an improper fraction.

26. $5\frac{1}{2}$

27. $3\frac{2}{3}$

28. $7\frac{1}{4}$

List the first five non-zero multiples for each number.

29. 5

30. 7

31. 8

Tic-Tac-Toe ~ Interview

Interior decorating, engineering, carpet laying, carpentry, architecture, tailoring and plumbing are a few occupations that use fractions regularly. Choose an occupation that uses fractions regularly (it can be one of those above or one approved by your teacher).

Step 1: Write interview questions for someone in that job. Consider what they do on a daily basis and how fractions are used.

Step 2: Interview someone who works in the occupation you chose. Record their responses to your questions.

Step 3: Write a one-page report to inform others about this occupation and how fractions are a necessary aspect of this job. Include the interview questions and the person's responses with your report.

Tic-Tac-Toe ~ Landscaping

Step 1: Draw your own backyard landscape with flower beds, or design one. Write the measurements, to the nearest inch, on all sides of each flower bed. Measure your own back yard or check your design measurements with an adult to make sure the measurements are realistic.

Step 2: Use advertisements, the internet or information from a home or garden store to choose an edging block to outline each flower bed. Record a drawing of the edging block and the real length in inches.

Step 3: How many edging blocks do you need to edge the perimeter of each flower bed in your landscape design?

Step 4: How many edging blocks do you need altogether?

Step 5: What is the total cost for the edging blocks?

Step 6: Display the landscape on a poster. Include the source of your materials along with Steps 1-5 on your poster. Show your calculations on a separate sheet of paper.

ADDING AND SUBTRACTING FRACTIONS

LESSON 4.2

 Find sums and differences of fraction expressions.

Sometimes when you add or subtract fractions the denominators are the same. Sometimes you may find that the fractions have different denominators. In this lesson you will learn how to deal with both situations.

EXPLORE!   **PIZZA PARTY!**

Janice was ordering pizzas for her friends. The table shows the fractions of the pizzas each person said they could eat.

	Cheese	Hawaiian	Pepperoni
Janice	$\frac{3}{10}$	$\frac{1}{6}$	$\frac{1}{4}$
Lakelynn	$\frac{1}{5}$	$\frac{1}{3}$	$\frac{3}{8}$
Alvaro	$\frac{3}{5}$	$\frac{1}{2}$	$\frac{5}{8}$
Jory	$\frac{1}{10}$	$\frac{2}{3}$	$\frac{1}{2}$

How much cheese pizza do Lakelynn and Alvaro think they can eat together? Lay out the fraction tiles to represent the amounts of cheese pizza which Lakelynn and Alvaro want. Combine the fraction tiles to find the total. Simplify the fraction, if needed. Draw a picture of the fraction model you used and write the addition equation on paper.

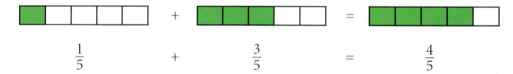

$$\frac{1}{5} \qquad + \qquad \frac{3}{5} \qquad = \qquad \frac{4}{5}$$

Step 1: Follow the process shown above to answer the following questions.
  a. How much cheese pizza will Janice and Jory eat together?
  b. How much Hawaiian pizza will Jory and Lakelynn eat together?
  c. How much pepperoni pizza will Lakelynn and Alvaro eat together?

Step 2: Look at the equations. What do you notice about the denominators of the fractions being added and the denominator of the answer? What about the numerators? How might you add fractions with common denominators without using fraction tiles?

How much Hawaiian pizza will Jory and Alvaro eat? Lay out fraction tiles to represent the amount of Hawaiian pizza Jory and Alvaro eat.

$$\frac{2}{3} \quad + \quad \frac{1}{2}$$

Find the least common denominator for the two fractions.

2: 2, 4, ⑥, 8

3: 3, ⑥, 9, 12   LCD = 6

Replace the original fraction tiles with equivalent fractions using the LCD. Combine the tiles to find the answer.

$$\frac{4}{6} \quad + \quad \frac{3}{6} \quad = \quad \frac{7}{6}$$

Simplify the fraction, if needed. All improper fractions should be written as mixed numbers.

$$\frac{7}{6} = 1\frac{1}{6}$$

Step 3: Follow the process shown above to answer the following questions. Draw pictures of the fraction tiles used in the process. Write the addition equation on your paper.

  a. How much cheese pizza will Janice and Lakelynn eat?
  b. How much Hawaiian pizza will Lakelynn and Alvaro eat?
  c. How much Hawaiian pizza will Janice and Jory eat?
  d. How much pepperoni pizza will Janice and Alvaro eat?
  e. How much pepperoni pizza will Janice and Jory eat?
  f. How much pepperoni pizza will Alvaro and Jory eat?

Step 4: How might you add fractions with unlike denominators without using fraction tiles?

ADD OR SUBTRACT FRACTIONS WITH COMMON DENOMINATORS

1. Add or subtract the numerators.
2. Write the sum or difference over the common denominator.
3. Simplify the fraction.

ADD OR SUBTRACT FRACTIONS WITH UNLIKE DENOMINATORS

1. Rewrite the fractions using the least common denominator (LCD).
2. Add or subtract the numerators.
3. Write the sum or difference over the common denominator.
4. Simplify the fraction.

After adding, subtracting, multiplying or dividing fractions, you must simplify the fraction. This means writing the fraction in simplest form, then changing improper fractions to mixed numbers if necessary.

EXAMPLE 1

Kiley practiced $\frac{1}{8}$ of her piano music in the morning. That afternoon she practiced $\frac{3}{8}$ of her music. What fraction of her piano music did she practice?

SOLUTION

Write the problem.

$$\frac{1}{8} + \frac{3}{8}$$

Add the numerators and write the sum over the common denominator.

$$\frac{1+3}{8} = \frac{4}{8}$$

Simplify the fraction.

$$\frac{4}{8} = \frac{1}{2}$$

Kiley practiced $\frac{1}{2}$ of her piano music.

EXAMPLE 2

Find the value of $\dfrac{5}{6} - \dfrac{1}{6}$.

SOLUTION

Subtract the numerators. Write the difference over the common denominator.

$$\frac{5-1}{6} = \frac{4}{6}$$

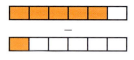

Simplify the fraction.

$$\frac{4}{6} = \frac{2}{3}$$

$$\frac{5}{6} - \frac{1}{6} = \frac{2}{3}$$

EXAMPLE 3

Mallory made $\frac{3}{4}$ gallon of ice cream. Logan made $\frac{1}{2}$ gallon. How many total gallons of ice cream do they have together?

SOLUTION

Write the problem.

$$\frac{3}{4} + \frac{1}{2}$$

Find the least common denominator for the set of fractions.

2: 2, ④, 6
4: ④, 8, 12

Rewrite the fractions using the LCD.

$$\frac{3}{4} = \frac{3}{4} \qquad \frac{1}{2} \overset{\times 2}{\underset{\times 2}{=}} \frac{2}{4}$$

Add the numerators.

$$\frac{3}{4} + \frac{2}{4} = \frac{3+2}{4} = \frac{5}{4}$$

Change the improper fraction to a mixed number.

$$\frac{5}{4} = 1\frac{1}{4}$$

Mallory and Logan made a total of $1\frac{1}{4}$ gallons of ice cream.

EXAMPLE 4

Find the value of $\frac{2}{3} - \frac{1}{4}$.

SOLUTION

Find the LCD.

3: 3, 6, 9, (12)
4: 4, 8, (12) 16

Rewrite the fraction using the LCD.

$$\frac{2}{3} \xrightarrow{\times 4} = \frac{8}{12} \xleftarrow{\times 4}$$

$$\frac{1}{4} \xrightarrow{\times 3} = \frac{3}{12} \xleftarrow{\times 3}$$

Subtract the numerators.

$$\frac{8}{12} - \frac{3}{12} = \frac{8-3}{12} = \frac{5}{12}$$

$$\frac{2}{3} - \frac{1}{4} = \frac{5}{12}$$

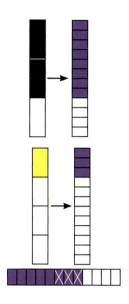

EXERCISES

Find each sum or difference. Write in simplest form.

1. $\frac{1}{4} + \frac{1}{4}$

2. $\frac{3}{5} + \frac{1}{5}$

3. $\frac{1}{8} + \frac{5}{8}$

4. $\frac{1}{10} + \frac{5}{10}$

5. $\frac{7}{9} + \frac{5}{9}$

6. $\frac{5}{6} - \frac{3}{6}$

7. $\frac{7}{8} - \frac{5}{8}$

8. $\frac{9}{15} - \frac{4}{15}$

9. $\frac{9}{11} - \frac{3}{11}$

10. Isaiah surveyed his class to see whether they thought the beaver was an appropriate mascot for their school. Four-eighths of the class thought it was a good choice. Three-eighths of the class thought another animal should be chosen as the mascot. One-eighth of his class did not want to answer the survey. What fraction of his class answered the survey?

11. Jace ran $\frac{3}{10}$ mile on Monday. On Tuesday he ran $\frac{5}{10}$ mile. How much further did he run on Tuesday than Monday?

Find each sum or difference. Write in simplest form.

12. $\frac{1}{8} + \frac{1}{2}$

13. $\frac{5}{12} + \frac{4}{6}$

14. $\frac{3}{10} + \frac{2}{5}$

15. $\frac{1}{2} + \frac{2}{6}$

16. $\frac{3}{4} + \frac{2}{3}$

17. $\frac{2}{3} - \frac{1}{2}$

18. $\frac{5}{6} - \frac{2}{8}$

19. $\frac{5}{9} - \frac{1}{3}$

20. $\frac{11}{12} - \frac{3}{4}$

21. Natasha's desk measured $\frac{3}{4}$ meter wide. Neil's desk was $\frac{2}{3}$ meter wide. How much wider was Natasha's desk than Neil's desk?

22. Lara ate $\frac{1}{2}$ of a sub sandwich. Lucas ate $\frac{3}{4}$ of a sub sandwich. How much did they eat altogether?

23. One-third of the students Ms. Jarrett teaches are sixth graders. Five-twelfths of the students she teaches are seventh graders. The rest are eighth graders.
    a. What fraction of her students are 6th and 7th graders?
    b. What fraction of her students are 8th graders?

24. Aaron rode his bike off a jump $\frac{7}{12}$ yard tall. The next day he rode his bike off a jump $\frac{8}{9}$ yard tall. How much taller was the second jump than the first jump?

REVIEW

Draw a line with the given measure.

25. $2\frac{1}{8}$ *in*

26. $4\frac{3}{4}$ *in*

27. $1\frac{1}{2}$ *in*

Estimate each difference.

28. $\frac{9}{10} - \frac{3}{8}$

29. $\frac{11}{12} - \frac{1}{5}$

30. $5\frac{1}{8} - 2\frac{1}{4}$

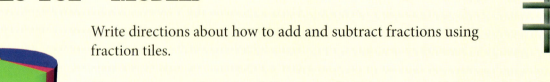

TIC-TAC-TOE ~ MODELS

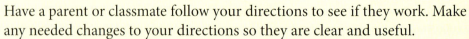

Write directions about how to add and subtract fractions using fraction tiles.

Have a parent or classmate follow your directions to see if they work. Make any needed changes to your directions so they are clear and useful.

Create a brochure using your directions to teach students how to use fraction tiles to show addition or subtraction of fractions.

ADDING AND SUBTRACTING MIXED NUMBERS

 Find sums and differences of expressions that include mixed numbers.

Mixed numbers are used in many real-world situations. Measurements are often given as mixed numbers. Amounts of each ingredient in recipes are also in mixed number form. You will need to add or subtract mixed numbers to solve problems in many situations.

EXPLORE!   **MIXING PAINT**

Kazi painted his bedroom. He mixed different colors of paint together to make new and exciting shades.

Step 1: Kazi purchased two small cans of paint for the ceiling. The can of red paint contained $2\frac{3}{4}$ pints. The can of blue paint contained $1\frac{5}{8}$ pints. He mixed the two cans of paint together. How much paint does he have for the ceiling?

    a. Write the problem as a mathematical expression.
    b. Write each mixed number as an improper fraction.
    c. Rewrite the fractions using the least common denominator.
    d. Find the sum or difference. Simplify your answer and convert it to a mixed number, if needed.

Step 2: Kazi bought $1\frac{1}{6}$ gallons of yellow and $4\frac{1}{3}$ gallons of green paint for the walls. He mixed the two colors of paint together. How much paint does he have available for the walls?

    a. Write the problem as a mathematical expression.
    b. Write each mixed number as an improper fraction.
    c. Rewrite the fractions using the least common denominator.
    d. Find the sum or difference. Simplify your answer and convert it to a mixed number, if needed.

Step 3: Kazi bought $2\frac{1}{5}$ pints of white paint for the trim. When he finished painting he still had $\frac{7}{10}$ pint of white paint left. How much white paint did Kazi use?

    a. Write the problem as a mathematical expression.
    b. Write each mixed number as an improper fraction.
    c. Rewrite the fractions using the least common denominator.
    d. Find the sum or difference. Simplify your answer and convert it to a mixed number, if needed.

Step 4: Explain in your own words how to add or subtract mixed numbers using the method in this **Explore!**

ADDING OR SUBTRACTING MIXED NUMBERS USING IMPROPER FRACTIONS

1. Write the mixed numbers as improper fractions.
2. If the denominators are different, rewrite the fractions using the least common denominator (LCD).
3. Add or subtract the numerators.
4. Write the sum or difference over the common denominator.
5. Simplify the fraction.

EXAMPLE 1

Find the value of $3\frac{1}{4} + 1\frac{2}{3}$.

SOLUTION

Change each mixed number to an improper fraction.

$$3\frac{1}{4} = \frac{13}{4} \qquad 1\frac{2}{3} = \frac{5}{3}$$

Write equivalent fractions with the least common denominator, 12.

$$\overset{\times 3}{\frac{13}{4}} = \frac{39}{12} \underset{\times 3}{} \qquad \text{and} \qquad \overset{\times 4}{\frac{5}{3}} = \frac{20}{12} \underset{\times 4}{}$$

Add the numerators.

$$\frac{39}{12} + \frac{20}{12} = \frac{39 + 20}{12} = \frac{59}{12}$$

Write as a mixed number.

$$\frac{59}{12} = 4\frac{11}{12}$$

$$3\frac{1}{4} + 1\frac{2}{3} = 4\frac{11}{12}$$

EXAMPLE 2

Seth ran a mile around the track. It took him $6\frac{5}{6}$ minutes to run a mile. His friend, Tremaine, also ran a mile on the track. It took Tremaine $7\frac{1}{3}$ minutes to run the mile. How much faster did Seth run than Tremaine?

SOLUTION

Write the problem.

$$7\frac{1}{3} - 6\frac{5}{6}$$

Write the mixed numbers as improper fractions.

$$7\frac{1}{3} = \frac{22}{3} \qquad 6\frac{5}{6} = \frac{41}{6}$$

Write equivalent fractions with the LCD of 6.

$$\frac{22}{3} = \frac{44}{6}$$

Subtract the numerators.

$$\frac{44}{6} - \frac{41}{6} = \frac{44 - 41}{6} = \frac{3}{6}$$

Write the fraction in simplest form.

$$\frac{3}{6} = \frac{1}{2}$$

Seth ran the mile $\frac{1}{2}$ minute faster than Tremaine.

EXERCISES

Find each sum. Write in simplest form.

1. $5\frac{1}{2} + 3\frac{1}{2}$

2. $4\frac{2}{3} + 1\frac{1}{6}$

3. $1\frac{1}{4} + 2\frac{1}{2}$

4. $4\frac{2}{9} + 2\frac{2}{3}$

5. $3\frac{2}{5} + 2\frac{3}{10}$

6. $1\frac{1}{5} + 4\frac{2}{3}$

7. $3\frac{1}{6} + 2\frac{3}{4}$

8. $1\frac{2}{3} + 1\frac{1}{2}$

9. $3\frac{1}{3} + 1\frac{1}{4}$

10. Silas went to the grocery store with his parents. They bought two kilograms of carrots. One kilogram is about $2\frac{1}{5}$ pounds. Silas wanted to find the number of pounds in two kilograms so he added $2\frac{1}{5} + 2\frac{1}{5}$. How many pounds are equal to two kilograms of carrots?

11. Cory nailed two boards together. The first one was $1\frac{5}{6}$ inches thick. The second one was $3\frac{1}{2}$ inches thick. How thick were the two boards together?

12. Natalie poured $7\frac{3}{8}$ ounces of club soda in a glass. She added $2\frac{1}{4}$ ounces of raspberry flavoring to the club soda. How much liquid was in the glass?

Find each difference. Write in simplest form.

13. $3\frac{7}{8} - 2\frac{3}{8}$

14. $2\frac{1}{6} - 1\frac{1}{3}$

15. $3\frac{3}{4} - \frac{1}{4}$

16. $4\frac{3}{7} - 1\frac{3}{14}$

17. $2\frac{4}{5} - 1\frac{1}{15}$

18. $5\frac{3}{4} - 1\frac{11}{12}$

19. $5\frac{5}{6} - 3\frac{1}{5}$

20. $4\frac{1}{5} - 2\frac{1}{6}$

21. $3\frac{5}{12} - 2\frac{3}{4}$

22. Nate made $2\frac{2}{3}$ quarts of salsa last year. This year he made $6\frac{5}{6}$ quarts of salsa. How much more salsa did he make this year?

23. Consuela spent $2\frac{1}{2}$ hours babysitting for her neighbor on Saturday. The next week she babysat for $3\frac{1}{4}$ hours. How much longer did she babysit the second week?

24. Sari stood $5\frac{1}{3}$ ft tall. Her mom stood $4\frac{5}{6}$ ft tall. How much taller was Sari than her mom?

25. Two African elephants weigh different amounts. The older one weighs $6\frac{5}{8}$ tons. His son weighs $3\frac{17}{40}$ tons. How much more does the older elephant weigh than the younger elephant?

Find the value of each expression. Write in simplest form.

26. $\dfrac{1}{2} + \dfrac{3}{8}$

27. $\dfrac{3}{4} + \dfrac{1}{5}$

28. $\dfrac{5}{6} + \dfrac{3}{5}$

29. $\dfrac{9}{10} - \dfrac{3}{5}$

30. $\dfrac{2}{3} - \dfrac{1}{4}$

31. $\dfrac{11}{12} - \dfrac{1}{5}$

TIC-TAC-TOE ~ YOU ARE THE AUTHOR

There are many children's books which include fractions, such as:

The Wishing Club: A Story about Fractions by Donna Jo Napoli
Give Me Half! by Stuart J. Murphy
Fraction Fun by David A. Adler & Nancy Tobin
Working With Fractions by David A. Adler & Edward Miller
Fraction Action by Loreen Leedy
Hershey's Fractions by Jerry Pallotta & Robert C. Bolster
Apple Fractions by Jerry Pallotta & Rob Bolster

Step 1: Read at least two of the books cited above. Write a paragraph summarizing the plot of each book that you read.

Step 2: Create a children's book which includes the concept of fractions. The story should be appropriate for children. Create a cover and illustrations for your story.

TIC-TAC-TOE ~ LIGHTS, CAMERA, ACTION

Adding and subtracting fractions occurs everyday.

Step 1: Make a list of situations where adding or subtracting fractions are used. Include problems that can be solved using addition or subtraction of fractions.

Step 2: Write a skit with two or more characters. The plot should have a problem and a solution using addition or subtraction of fractions.

Script Example: Reid: *write what he says.*
Kellene: *write what she says.*
Include any actions they are to take in parentheses, like (walk across stage) or (face each other).

TIC-TAC-TOE ~ TRIATHLON

A triathlon is an endurance event that consists of swimming, cycling and running. The length of each segment varies depending on the triathlon. Answer the questions below about participants in different triathlons.

1. Pacific Olympic Triathlon

   Swim: $\frac{15}{16}$ miles   Cycle: 28 miles   Run: $6\frac{1}{5}$ miles

   a. How long is the entire triathlon?
   b. Marita has $2\frac{1}{8}$ miles left in the running segment. How far has she traveled so far?

2. Sprint Triathlon

   Swim: $\frac{15}{32}$ mile   Cycle: $9\frac{3}{5}$ miles   Run: $3\frac{1}{10}$ miles

   a. How long is the entire triathlon?
   b. Quan has finished the swimming segment and has biked 6 miles. How far does he still need to go to finish the triathlon?

3. Snake Falls Triathlon

   Swim: $\frac{3}{5}$ mile   Cycle: 13 miles   Run: $3\frac{1}{10}$ miles

   a. How long is the entire triathlon?
   b. Jerome has $2\frac{1}{8}$ miles left in the cycling segment. How many miles is he from the finish line?

4. Equinox Triathlon

   Swim: $\frac{3}{5}$ mile   Cycle: $15\frac{1}{2}$ miles   Run: $6\frac{1}{5}$ miles

   a. How long is the entire triathlon?
   b. LeAnne has $1\frac{1}{4}$ miles left in the cycling segment. How far has she traveled from the starting line of the triathlon?

5. Gator Triathlon

   Swim: $\frac{25}{88}$ mile   Cycle: 12 miles   Run: $3\frac{1}{10}$ miles

   a. How long is the entire triathlon?
   b. Carlos has almost finished the swimming segment. He had completed $\frac{21}{88}$ mile so far. How far does he still need to go to finish the swimming segment?
   c. How far does Carlos need to go to finish the triathlon?

ADDING AND SUBTRACTING BY RENAMING

LESSON 4.4

Calculate sums and differences of mixed numbers by renaming.

Y ou learned one method for adding and subtracting mixed numbers in **Lesson 4.3**. You converted each mixed number into an improper fraction and followed the process for adding or subtracting two fractions. Another method for adding or subtracting mixed numbers is called renaming. Renaming can also be called borrowing or regrouping.

Paula bought a turkey for a family dinner. It weighed $10\frac{3}{4}$ pounds. Her mother did not realize Paula had bought a turkey. Her mother also bought a turkey that weighed $12\frac{2}{3}$ pounds. How many total pounds of turkey did they have for the family dinner?

Write the problem.	$10\frac{3}{4} + 12\frac{2}{3}$
Rewrite the fractions using the LCD, 12.	$10\frac{9}{12} + 12\frac{8}{12}$
Add each part of the mixed numbers.	$\begin{array}{r} 10\frac{9}{12} \\ + 12\frac{8}{12} \\ \hline 22\frac{17}{12} \end{array}$
Rename the improper fraction as a mixed number.	$\frac{17}{12} = 1\frac{5}{12}$
Add the sum of the whole numbers to the renamed fraction.	$22 + 1\frac{5}{12} = 23\frac{5}{12}$

Paula's family had a total of $23\frac{5}{12}$ pounds of turkey for dinner.

How much larger was the turkey Paula's mother bought than the one Paula had purchased?

Write the problem.	$12\frac{2}{3} - 10\frac{3}{4}$
Rewrite the fractions using the LCD, 12.	$12\frac{8}{12} - 10\frac{9}{12}$
Rename $12\frac{8}{12}$ as $11\frac{12}{12} + \frac{8}{12}$ which is $11\frac{20}{12}$.	$\begin{array}{r} 11\frac{20}{12} \\ - 10\frac{9}{12} \\ \hline 1\frac{11}{12} \end{array}$
Subtract each part of the mixed numbers.	

> If the second fraction is larger than the first fraction, you must rename the first fraction. Borrow 1 from 12 and rename as $\frac{12}{12}$.

The turkey Paula's mother bought was $1\frac{11}{12}$ pounds larger than Paula's turkey.

There are two situations where renaming mixed numbers to find the sum or difference is needed.

- Rename when subtracting if the fraction in the first number is smaller than the fraction in the second number.
- Rename when adding if the two fractions add to more than 1.

EXAMPLE 1

Find the value of $4\frac{3}{5} + 2\frac{7}{10}$.

SOLUTION

Rewrite the fractions using the LCD, 10.

$$4\frac{6}{10} + 2\frac{7}{10}$$

Add the fractions and the ~~whole~~ numbers.

$$\begin{array}{r} 4\frac{6}{10} \\ + 2\frac{7}{10} \\ \hline 6\frac{13}{10} \end{array}$$

Rename the improper fraction.

$$6\frac{13}{10} = 6 + 1\frac{3}{10} = 7\frac{3}{10}$$

$$4\frac{3}{5} + 2\frac{7}{10} = 7\frac{3}{10}$$

EXAMPLE 2

Mahavir biked $14\frac{5}{6}$ miles on Saturday. He biked $18\frac{1}{4}$ miles on Sunday. How much further did he bike on Sunday?

SOLUTION

Write the problem.

$$18\frac{1}{4} - 14\frac{5}{6}$$

Rewrite using the LCD, 12.

$$18\frac{3}{12} - 14\frac{10}{12}$$

Rename the first fraction because $\frac{3}{12}$ is smaller than $\frac{10}{12}$.

$$\begin{array}{r} 17\frac{15}{12} \\ - 14\frac{10}{12} \\ \hline 3\frac{5}{12} \end{array}$$

Rename $18\frac{3}{12}$ as $17\frac{15}{12}$.

Mahavir rode $3\frac{5}{12}$ miles further on Sunday.

EXAMPLE 3

The longest python on record was 33 feet long. The average male python grows to be $18\frac{1}{6}$ feet long. What is the difference between the average male python and the world's longest python?

SOLUTION

Write the problem. $33 - 18\frac{1}{6}$

Rename 33 as $32\frac{6}{6}$.

$$\begin{array}{r} 32\frac{6}{6} \\ -\ 18\frac{1}{6} \\ \hline 14\frac{5}{6} \end{array}$$

The world's longest python was $14\frac{5}{6}$ feet longer than the average male python.

EXERCISES

1. Explain one type of problem when renaming should be used to add or subtract mixed numbers.

2. Copy and complete each set of equivalent mixed numbers.
   a. $4\frac{1}{4} = \underline{\quad}\frac{5}{4}$ **b.** $2\frac{2}{5} = 1\frac{}{5}$ **c.** $6\frac{7}{4} = \underline{\quad}\frac{3}{4}$

Find each sum or difference. Write in simplest form.

3. $2\frac{3}{5} + 4\frac{4}{5}$

4. $6\frac{2}{3} + 5\frac{5}{6}$

5. $9\frac{7}{8} - 1\frac{3}{8}$

6. $10\frac{3}{7} - 7\frac{5}{7}$

7. $8\frac{1}{2} + 3\frac{3}{4}$

8. $15\frac{1}{3} - 10\frac{1}{6}$

9. $3 - 1\frac{3}{10}$

10. $4\frac{11}{12} + 1\frac{1}{3}$

11. $8\frac{2}{3} - 7\frac{5}{6}$

12. $22\frac{1}{2} + 10\frac{5}{7}$

13. $7 - 3\frac{5}{8}$

14. $6\frac{3}{5} - 1\frac{1}{6}$

15. A junior-sized football is $10\frac{3}{8}$ inches long and $5\frac{7}{8}$ inches wide. What is the difference between the football's length and width?

16. Kirk caught a fish that weighed $2\frac{9}{16}$ pounds. His little brother caught a fish that weighed $3\frac{3}{4}$ pounds. What was the total weight of the fish the boys caught?

17. Owen walked $1\frac{4}{5}$ miles on Tuesday and $3\frac{1}{2}$ miles on Wednesday. How far did he walk in the two days combined?

18. A bag contained 2 cups of raisins. Five-eighths of a cup of raisins was used in a recipe. How many cups of raisins remain in the bag?

19. Alberto and Marco wrote songs for their band. Alberto's song was $2\frac{1}{2}$ minutes long. Marco's song was $4\frac{1}{6}$ minutes long. How much longer was Marco's song than Alberto's song?

20. How do you know when you need to rename one of the numbers when subtracting mixed numbers?

21. The rim on a basketball hoop is 10 feet off the ground. Ron jumped and reached $8\frac{5}{8}$ feet off the ground. How much higher would Ron need to jump to touch the rim?

REVIEW

22. George had a board that was 5.5 meters long. He cut it into 4 equal pieces. How long was each piece?

23. Mahei earned $33.95 in tips. Lei earned 2.4 times that amount in tips. How much did Lei earn in tips?

24. Janell ran a 26.2 mile marathon. She ran each mile in 11.5 minutes. Assuming she ran every mile equally, how long did it take her to run the marathon?

TIC-TAC-TOE ~ CREATE THE PROBLEM

Instead of finding the sum of an addition problem, you must find two addends that equal a given sum.

Write two addition problems that equal the given sum. At least one problem in each set must have two addends with unlike denominators.

Example: $? + ? = \frac{1}{2}$

$\frac{1}{8} + \frac{3}{8} = \frac{4}{8} = \frac{1}{2}$ OR $\frac{1}{3} + \frac{1}{6} = \frac{2}{6} + \frac{1}{6} = \frac{3}{6} = \frac{1}{2}$

The sums are:

1. $\frac{1}{4}$ **2.** $\frac{1}{3}$ **3.** $\frac{3}{7}$ **4.** $\frac{5}{18}$ **5.** $\frac{3}{4}$ **6.** $\frac{7}{8}$

7. $\frac{11}{14}$ **8.** $\frac{6}{13}$ **9.** $\frac{7}{15}$ **10.** $\frac{5}{6}$ **11.** $\frac{7}{12}$ **12.** $\frac{2}{5}$

PERIMETER WITH FRACTIONS

Add lengths, including fractions and mixed numbers, of the sides of polygons to find perimeters.

If you walk around a football field or a city block, you have traveled the perimeter of something. **Perimeter** is the distance around a closed figure. When you walk all the way around a football field, you have walked its perimeter. When you travel down each side of a city block and back to your starting point, you have traveled the perimeter of the block.

A **polygon** is a closed figure formed by three or more line segments. To find the perimeter of any given polygon you add the lengths of all sides.

FINDING PERIMETER

1. Measure all sides (if necessary).
2. Add the lengths of all sides together.

EXAMPLE 1

Measure each side of the rectangle in inches. Find the perimeter of the rectangle.

SOLUTION

Use a ruler to measure each side of the shape to the nearest sixteenth of an inch.

Length = $\frac{3}{4}$ in   Width = $\frac{1}{4}$ in

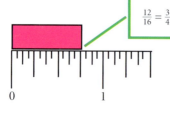

$$\frac{12}{16} = \frac{3}{4}$$

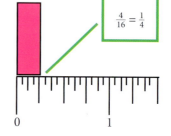

$$\frac{4}{16} = \frac{1}{4}$$

Add all four sides of the rectangle together. $\quad \frac{3}{4} + \frac{1}{4} + \frac{3}{4} + \frac{1}{4} = \frac{8}{4}$

Simplify. $\quad\quad\quad\quad\quad\quad\quad\quad\quad\quad \frac{8}{4} = 2$

The perimeter of the rectangle is 2 inches.

Opposite sides of rectangles are the same length.

EXAMPLE 2

Use the given measurements to find the perimeter of the polygon.

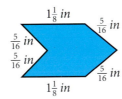

SOLUTION

Add the lengths of all sides together.

$$1\frac{1}{8} + \frac{5}{16} + \frac{5}{16} + 1\frac{1}{8} + \frac{5}{16} + \frac{5}{16}$$

Change the mixed numbers to improper fractions. $1\frac{1}{8} = \frac{9}{8}$

Use the LCD to rename the fractions.
The LCD is 16.

$$\frac{9}{8} = \frac{18}{16}$$

Add the sides of the polygon.

$$\frac{18}{16} + \frac{5}{16} + \frac{5}{16} + \frac{18}{16} + \frac{5}{16} + \frac{5}{16} = \frac{56}{16}$$

Simplify.

$$\frac{56}{16} = \frac{7}{2} = 3\frac{1}{2} \text{ or } \frac{56}{16} = 3\frac{8}{16} = 3\frac{1}{2}$$

The perimeter of the polygon is $3\frac{1}{2}$ inches.

EXAMPLE 3

Brayden went for a walk. He walked around a rectangular city block. First he walked $66\frac{2}{3}$ yards. He turned right and walked $76\frac{1}{6}$ yards. He turned right two more times and ended up where he started. What was the perimeter of the city block?

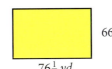

SOLUTION

Brayden walked in a rectangular pattern.
Add the four sides.

$$66\frac{2}{3} + 76\frac{1}{6} + 66\frac{2}{3} + 76\frac{1}{6} = ? \text{ yards}$$

Use the LCD to rename the fractions.
The LCD is 6.

$$66\frac{2}{3} = 66\frac{4}{6}$$

Add the sides of the city block.

$$66\frac{4}{6} + 76\frac{1}{6} + 66\frac{4}{6} + 76\frac{1}{6} = 284\frac{10}{6} \text{ yards}$$

Simplify.

$$\frac{10}{6} = 1\frac{4}{6} \rightarrow 284 + 1 + \frac{4}{6} = 285\frac{4}{6} = 285\frac{2}{3}$$

The perimeter of the city block Brayden walked was $285\frac{2}{3}$ yards.

EXAMPLE 4 | Use the given measurement to find the perimeter of the square.

SOLUTION

Add all four sides of the square together. $1\frac{1}{2} + 1\frac{1}{2} + 1\frac{1}{2} + 1\frac{1}{2} = 4\frac{4}{2}$

Simplify. $4\frac{4}{2} = 4 + 2 = 6$

The perimeter of the square is 6 inches.

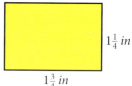

$1\frac{1}{2}$ in

> All sides of a square are the same length.

EXERCISES

Use the given measurements to find the perimeter of each rectangle. Write in simplest form.

1. 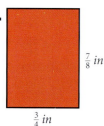 $\frac{1}{8}$ in

$\frac{5}{8}$ in

2. 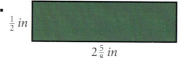 $\frac{13}{16}$ in

$\frac{7}{16}$ in

3. $1\frac{1}{4}$ in

$1\frac{3}{4}$ in

4. $\frac{7}{8}$ in

$\frac{3}{4}$ in

5. 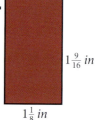 $\frac{1}{2}$ in

$2\frac{5}{8}$ in

6. $1\frac{9}{16}$ in

$1\frac{1}{8}$ in

7. Measure and record the lengths of the sides of your desk or a table to the nearest quarter inch. Find the perimeter.

8. Measure and record the lengths of the sides of a piece of notebook paper to the nearest sixteenth inch. Find the perimeter.

Use the given measurement to find the perimeter of each square. Write in simplest form.

9. $\frac{11}{16}$ in

10. $\frac{7}{8}$ in

11. 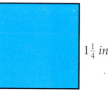 $1\frac{1}{4}$ in

Measure one side of each square to the nearest sixteenth of an inch using a customary ruler. Find each perimeter. Write in simplest form.

12.

13.

14.

Find the perimeter of each polygon. Write in simplest form.

15.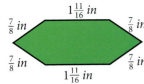
$\frac{7}{8}$ in, $1\frac{11}{16}$ in, $\frac{7}{8}$ in, $\frac{7}{8}$ in, $1\frac{11}{16}$ in, $\frac{7}{8}$ in

16.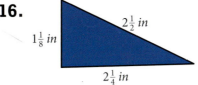
$1\frac{1}{8}$ in, $2\frac{1}{2}$ in, $2\frac{1}{4}$ in

17.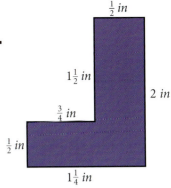
$\frac{1}{2}$ in, $1\frac{1}{2}$ in, 2 in, $\frac{3}{4}$ in, $\frac{1}{2}$ in, $1\frac{1}{4}$ in

18. Jenna's room was a perfect square. The length of one wall measured $15\frac{1}{6}$ feet. What is the perimeter of Jenna's room?

19. Thi walked $20\frac{1}{8}$ yards toward the office. She turned right and walked $35\frac{3}{4}$ yards toward the parking lot. She walked in a rectangle until she was back where she started. What is the perimeter of the rectangle Thi walked?

REVIEW

Find the value of each expression. Write in simplest form.

20. $9\frac{1}{4} - 3\frac{3}{4}$

21. $15\frac{1}{6} - 9\frac{5}{6}$

22. $40\frac{1}{2} - 22\frac{2}{3}$

23. $2\frac{1}{3} + 3\frac{3}{4}$

24. $9\frac{4}{5} + 3\frac{1}{6}$

25. $2\frac{1}{6} + 8\frac{5}{8}$

Simplify each fraction. Write as a mixed number.

26. $\frac{26}{10}$

27. $\frac{84}{9}$

28. $\frac{420}{105}$

TIC-TAC-TOE ~ MEASURE THE PERIMETER

Picture frames come in many sizes.

Step 1: Find ten or more different sized frames at home or at a store (they do not all have to be rectangular).

Step 2: Use a customary ruler or measuring tape to measure the lengths of the sides of each frame to the nearest sixteenth of an inch.

Step 3: Draw a sketch of the shape of each frame. Record the lengths of the sides you measured.

Step 4: Calculate the perimeter of each frame.

Vocabulary

perimeter
polygon

Estimate sums and differences of expressions with fractions and mixed numbers.
Find sums and differences of fraction expressions.
Find sums and differences of expressions that include mixed numbers.
Calculate sums and differences of mixed numbers by renaming.
Add lengths, including fractions and mixed numbers, of the sides of polygons to find perimeters.

Lesson 4.1 ~ Estimating Sums and Differences

Estimate each sum or difference. Round to 0, $\frac{1}{2}$ or 1 before adding or subtracting.

1. $\frac{1}{8} + \frac{2}{3}$

2. $\frac{4}{10} + \frac{7}{8}$

3. $\frac{8}{9} - \frac{2}{5}$

4. $\frac{4}{7} - \frac{1}{9}$

5. $\frac{9}{10} + \frac{1}{6}$

6. $\frac{11}{20} - \frac{8}{15}$

Estimate each sum or difference. Round to the nearest whole number before adding or subtracting.

7. $1\frac{7}{8} + 2\frac{8}{9}$

8. $6\frac{3}{5} + 2\frac{1}{8}$

9. $3\frac{2}{3} - 1\frac{6}{7}$

10. $4\frac{5}{13} - 2\frac{1}{6}$

11. $10\frac{1}{5} + 4\frac{1}{9}$

12. $12\frac{4}{5} - 1\frac{1}{6}$

Lesson 4.2 ~ Adding and Subtracting Fractions

Find each sum or difference. Write in simplest form.

13. $\frac{3}{8} + \frac{2}{8}$

14. $\frac{6}{7} + \frac{6}{7}$

15. $\frac{8}{9} - \frac{2}{9}$

16. $\frac{5}{6} - \frac{1}{6}$

17. $\frac{7}{10} + \frac{2}{5}$

18. $\frac{1}{6} + \frac{3}{4}$

19. $\frac{7}{9} - \frac{2}{3}$

20. $\frac{11}{12} - \frac{1}{4}$

21. $\frac{1}{9} + \frac{1}{2}$

22. Corrie had $\frac{1}{2}$ cup of brown sugar. She borrowed $\frac{1}{4}$ cup of brown sugar in order to have enough for her chocolate chip cookie recipe. How much brown sugar did the recipe call for?

23. Tonda's sunflower plant was $\frac{7}{8}$ yard tall. Jasmine's sunflower plant was $\frac{2}{3}$ yard tall. How much taller was Tonda's sunflower plant than Jasmine's sunflower plant?

Lesson 4.3 ~ Adding and Subtracting Mixed Numbers

Find each sum or difference. Write in simplest form.

24. $1\frac{1}{5} + 2\frac{4}{5}$

25. $2\frac{3}{4} + 1\frac{1}{2}$

26. $3\frac{5}{8} - 1\frac{1}{8}$

27. $5\frac{7}{8} - 2\frac{5}{8}$

28. $5\frac{1}{8} + 2\frac{3}{4}$

29. $4\frac{5}{6} + 2\frac{2}{3}$

30. $2\frac{2}{3} - 1\frac{3}{4}$

31. $5\frac{1}{2} - 3\frac{5}{9}$

32. $4\frac{2}{3} - 1\frac{1}{2}$

33. Travis used $4\frac{3}{4}$ quarts of oil when he changed his car's oil. A week later he checked his oil. He had to put $1\frac{1}{3}$ quarts of oil in the car because there was a leak. How much oil did Travis use in all?

34. Quinn's dad is $6\frac{1}{2}$ feet tall. Quinn is $5\frac{1}{4}$ feet tall. How much taller is Quinn's dad than Quinn?

Lesson 4.4 ~ Adding and Subtracting by Renaming

Find each sum or difference using renaming. Write in simplest form.

35. $3\frac{2}{7} + 1\frac{6}{7}$

36. $10 - 4\frac{8}{9}$

37. $13\frac{2}{5} - 9\frac{1}{2}$

38. $7\frac{5}{6} + 3\frac{3}{4}$

39. $5 - 3\frac{1}{3}$

40. $2\frac{4}{5} + 1\frac{7}{10}$

41. Rory made $8\frac{1}{2}$ quarts of punch for a birthday party. When the punch ran out, Rory had to make $2\frac{5}{8}$ more quarts. How much punch did Rory make altogether?

42. Thurston bought 50 pounds of grain for his cows. At the end of the week he had $7\frac{7}{8}$ pounds of grain left. How many pounds of grain did the cows eat?

Use the given measurements to find the perimeter of each polygon. Write in simplest form.

43.

SQUARE $2\frac{3}{4}$ in

44.

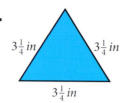

$3\frac{1}{4}$ in $3\frac{1}{4}$ in

$3\frac{1}{4}$ in

45.

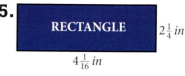

RECTANGLE $2\frac{1}{4}$ in

$4\frac{1}{16}$ in

Use a customary ruler to measure one side of each square to the nearest sixteenth of an inch. Find each perimeter. Write in simplest form.

46.

47.

48.

49. Omid ran around the perimeter of a basketball court. The width of the court was $50\frac{1}{4}$ feet and the length was $90\frac{1}{2}$ feet. He ended where he began. What is the distance that Omid ran?

 **TIC-TAC-TOE ~ FIND THE SUM**

$$\frac{1}{2} + \frac{1}{3} + \frac{1}{4}$$

There are times where you will be finding the sum of more than two fractions or mixed numbers. You will have to change all mixed numbers to improper fractions to add the multiple fractions. Then find the least common denominator (LCD) of the entire set of fractions.

Example: $1\frac{1}{2} + \frac{1}{4} + \frac{3}{8}$

Rewrite $1\frac{1}{2}$ as $\frac{3}{2}$

The LCD of $\frac{3}{2}$, $\frac{1}{4}$ and $\frac{3}{8}$ is 8.

Use the LCD for the set of fractions. Rewrite each fraction. $\frac{3}{2} = \frac{12}{8}, \frac{1}{4} = \frac{2}{8}, \frac{3}{8}$ remains $\frac{3}{8}$

Add the numerators over the common denominator. $\frac{12}{8} + \frac{2}{8} + \frac{3}{8} = \frac{12 + 2 + 3}{8} = \frac{17}{8}$

Simplify. Write improper fractions as mixed numbers. $\frac{17}{8} = 2\frac{1}{8}$

Write 20 different simplified fractions or mixed numbers on 3×5 cards. Draw three cards from the stack and write an addition equation using the three fractions. Use the process above to find the sum. Do this for at least ten problems.

CAREER FOCUS

KEN
DEPUTY SHERIFF

I am a deputy sheriff. I patrol roads, highways and business areas and enforce traffic and criminal laws. When there are accidents on our roads, I investigate them and make reports. I also conduct investigations and gather evidence from crime scenes. Part of conducting an investigation means taking statements from witnesses, as well as talking to people suspected of crimes. If needed, I even sometimes have to arrest and transport people suspected of crimes to jail or court. Most importantly, I inform the public about the law and answer questions about rules and regulations.

When investigating a crime, it is very important that the most accurate information is recorded. Math helps me to do this. During accident reconstructions I have to determine how fast people were going and in which angles they were headed. These mathematical values can reveal important information about what actually happened. I also use math in estimating my arrival time when I get called. I have to estimate how far away from a call I am and then calculate how long it will take me to get there if I travel at a certain speed. I also use simple math every morning when I test my RADAR for accuracy.

To become a deputy sheriff requires a high school diploma and at least two years of college. Usually people take classes in law enforcement, though sometimes experience in the field can substitute for coursework.

In the county where I am employed, starting salaries for deputy sheriffs range from $3,172 to $4,420 per month. Salaries vary depending upon where a person works and how much experience they have.

What I like most about being a deputy sheriff is helping people and trying to keep them safe. I really enjoy meeting people and presenting an image that results in a positive perception of my chosen profession.

CORE FOCUS ON DECIMALS & FRACTIONS
BLOCK 5 ~ MULTIPLYING AND DIVIDING FRACTIONS

LESSON 5.1   MULTIPLYING FRACTIONS WITH MODELS ----------------------------------- 136
                 EXPLORE! FRACTION ACTION

LESSON 5.2   MULTIPLYING FRACTIONS -- 139

LESSON 5.3   DIVIDING FRACTIONS WITH MODELS ------------------------------------ 143
                 EXPLORE! WHAT FITS?

LESSON 5.4   DIVIDING FRACTIONS --- 147

LESSON 5.5   ESTIMATING PRODUCTS AND QUOTIENTS ------------------------------- 151
                 EXPLORE! 4-H CLUB

LESSON 5.6   MULTIPLYING AND DIVIDING FRACTIONS AND WHOLE NUMBERS ----------- 155

LESSON 5.7   MULTIPLYING AND DIVIDING MIXED NUMBERS --------------------------- 159
                 EXPLORE! SCRAPBOOKING

REVIEW   BLOCK 5 ~ MULTIPLYING AND DIVIDING FRACTIONS ---------------------- 163

WORD WALL

RECIPROCAL

BLOCK 5 ~ MULTIPLYING AND DIVIDING FRACTIONS

TIC-TAC-TOE

ADVERTISEMENTS

Find advertisements that state "Buy One, Get One Half Off." Figure out the savings for purchases.

See page 162 for details.

MULTIPLICATION OF MULTIPLE FRACTIONS

Solve multiplication problems with three or more fractions.

See page 142 for details.

CHOOSE YOUR OWN ADVENTURE

Create a story. Character chooses between two different expressions with fractions to solve.

See page 142 for details.

CHANGING RECIPES

Find the amount of each ingredient to cut a recipe in half, make $1\frac{1}{2}$ times or triple it.

See page 162 for details.

FRACTION BINGO

Design a fraction BINGO game.

See page 154 for details.

CELL PHONE PLANS

Create a poster to compare the price-per-person and the minutes-per-person for family cell phone plans.

See page 150 for details.

HOW MANY MINUTES?

Find the number of minutes in a fraction of an hour.

See page 158 for details.

LEARNING WITH LYRICS

Write a song about fractions that teaches common errors when adding, subtracting, multiplying or dividing.

See page 162 for details.

TEACH ME!

Create a flap book to teach other students how to multiply and divide fractions using models.

See page 146 for details.

Use models to multiply fractions.

EXPLORE!   **FRACTION ACTION**

Find the value of $\frac{1}{6} \times \frac{3}{4}$ (which is read $\frac{1}{6}$ of $\frac{3}{4}$) by completing the following steps.

Step 1: Divide a piece of paper horizontally into as many sections as are shown in the denominator of one of the factors. For $\frac{3}{4}$, divide the paper into 4 horizontal sections.

Step 2: Use blue to color in as many horizontal sections as are shown in the numerator. For $\frac{3}{4}$, color in 3 of the 4 sections.

Step 3: Divide the piece of paper vertically into as many sections as are shown in the denominator of the factor. For $\frac{1}{6}$, divide the paper into 6 vertical sections.

Step 4: Use yellow to color in as many vertical sections as are shown in the numerator. For $\frac{1}{6}$, color in one of the 6 vertical sections.

Blue and yellow make green.

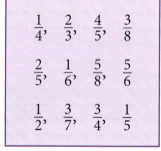

$$\frac{1}{4}, \quad \frac{2}{3}, \quad \frac{4}{5}, \quad \frac{3}{8}$$

$$\frac{2}{5}, \quad \frac{1}{6}, \quad \frac{5}{8}, \quad \frac{5}{6}$$

$$\frac{1}{2}, \quad \frac{3}{7}, \quad \frac{3}{4}, \quad \frac{1}{5}$$

Step 5: The answer to your product is $\dfrac{\text{number of sections shaded green}}{\text{total number of sections}}$.

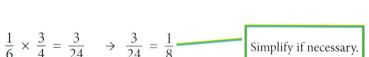

$$\frac{1}{6} \times \frac{3}{4} = \frac{3}{24} \quad \rightarrow \quad \frac{3}{24} = \frac{1}{8}$$

Simplify if necessary.

Step 6: Repeat **Steps 1-5** by finding the product of at least four different pairs of fractions from the purple box.

Step 7: Does it matter which fraction in the equation is shaded first? Explain your reasoning.

EXAMPLE 1

Rafael's mom had $\frac{3}{4}$ cup of blueberries. She split the blueberries among her three children. Each one received $\frac{1}{3}$ of the blueberries. What fraction of a cup of blueberries did each child receive?

SOLUTION

Write the problem.

$$\frac{1}{3} \times \frac{3}{4}$$

Draw a rectangle and divide it into fourths horizontally. Shade $\frac{3}{4}$ of the rectangle, or three of the four sections, with blue.

Divide the rectangle vertically into thirds. Shade $\frac{1}{3}$ of the rectangle, or one of the vertical sections, with yellow.

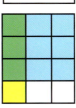

There are now 12 sections on the rectangle. This is the value of the denominator.

$$\frac{\square}{12}$$

There are 3 sections that were shaded twice. This is the value of the numerator. Simplify.

$$\frac{3}{12} = \frac{1}{4}$$

Each child received $\frac{1}{4}$ cup of blueberries.

EXAMPLE 2

Find the value of $\frac{4}{5} \times \frac{2}{3}$ using models.

SOLUTION

Draw a rectangle and divide it into thirds horizontally. Shade two of the three sections with blue.

Divide the rectangle vertically into fifths. Shade four of the five vertical sections with yellow.

There are 15 sections on the rectangle. This is the value of the denominator.

$$\frac{\square}{15}$$

There are 8 sections shaded twice. This is the value of the numerator.

$$\frac{8}{15}$$

The fraction is in simplest form, so $\frac{4}{5} \times \frac{2}{3} = \frac{8}{15}$.

1. Find the value of $\frac{3}{8} \times \frac{1}{2}$ by completing the following steps.

   a. Draw a rectangle.
   b. Divide the rectangle horizontally into as many sections as the denominator of the second fraction. Color in as many sections as the numerator of the second fraction.
   c. Divide the rectangle vertically into as many sections as the denominator of the first fraction. Color in as many sections as the numerator of the first fraction.
   d. Write a fraction where the total number of sections is the denominator and the total number of sections that are colored twice is the numerator.
   e. What is the answer in simplest form?

Use the procedure from Exercise 1 to find each product. Write the answer in simplest form.

2. $\frac{1}{4} \times \frac{3}{6}$ **3.** $\frac{5}{6} \times \frac{1}{2}$ **4.** $\frac{3}{5} \times \frac{1}{3}$

5. $\frac{1}{6} \times \frac{2}{5}$ **6.** $\frac{2}{4} \times \frac{4}{5}$ **7.** $\frac{1}{3} \times \frac{4}{6}$

Write the equation to match each of the following models. Write the answer in simplest form.

8. ___ × ___ = ___   **9.** ___ × ___ = ___   **10.** ___ × ___ = ___

11. A container held $\frac{1}{3}$ cup cream cheese. Becca put $\frac{1}{2}$ of the cream cheese on a bagel. What fraction of a cup of cream cheese did she put on the bagel?

12. About $\frac{7}{10}$ of the Earth's surface is water. The Pacific Ocean makes up $\frac{1}{2}$ of this water. What fraction of the Earth is covered by the Pacific Ocean?

REVIEW

Find each difference using renaming. Write in simplest form.

13. $4\frac{1}{3} - 2\frac{3}{4}$ **14.** $1\frac{2}{5} - \frac{3}{4}$ **15.** $6\frac{1}{6} - 4\frac{2}{3}$

Find each sum. Write in simplest form.

16. $1\frac{1}{2} + 2\frac{1}{6}$ **17.** $7\frac{1}{4} + 3\frac{2}{3}$ **18.** $5\frac{1}{3} + 4\frac{1}{4}$

MULTIPLYING FRACTIONS

 Find products of expressions involving two fractions.

Olivia's mom gave her $\frac{2}{3}$ of an hour to work on her homework with Tessa before they left for the grocery store. Olivia and Tessa used $\frac{1}{2}$ of that time for math homework. Olivia thought this equaled $\frac{1}{2}$ hour. Tessa said she was wrong. She said it equaled $\frac{1}{3}$ hour. Who was correct?

Olivia and Tessa wrote the problem as a mathematical expression.

$$\frac{1}{2} \text{ of } \frac{2}{3} \longrightarrow \frac{1}{2} \times \frac{2}{3}$$

MULTIPLYING FRACTIONS

For any numbers *a*, *b*, *c* and *d*:

$$\frac{a}{b} \times \frac{c}{d} = \frac{a \times c}{b \times d}$$

EXAMPLE 1 Find the value of $\frac{1}{2} \times \frac{2}{3}$.

SOLUTION

Find the products of the numerators and denominators.
Simplify.

$$\frac{1}{2} \times \frac{2}{3} = \frac{1}{3}$$

$$\frac{1}{2} \times \frac{2}{3} = \frac{1 \times 2}{2 \times 3} = \frac{2}{6}$$

$$\frac{2}{6} = \frac{1}{3}$$

When a numerator and denominator of either fraction have a common factor you can simplify before multiplying.

EXAMPLE 2 Find the value of $\frac{1}{8} \times \frac{4}{7}$.

SOLUTION

Find a common factor of one numerator and one denominator. The GCF of the numerator, 4, and denominator, 8, is 4.

Divide that numerator and denominator by 4. Write the factor above or below each simplified number in the fractions.

Multiply the numerators and denominators.

$$\frac{1}{8} \times \frac{4}{7} = \frac{1}{14}$$

> Simplifying before multiplying means the product will not need to be simplified.

$$\frac{1}{8} \times \frac{4}{7} = \frac{1 \times \overset{1}{\cancel{4}}}{\underset{2}{\cancel{8}} \times 7}$$

$$\frac{1 \times 1}{2 \times 7} = \frac{1}{14}$$

EXAMPLE 3

Kristi had $\frac{3}{8}$ cup of yogurt. She put $\frac{4}{9}$ of the yogurt into a smoothie. What fraction of a cup of yogurt did she use in the smoothie?

SOLUTION

Write the problem.

$$\frac{4}{9} \times \frac{3}{8}$$

The GCF of the numerator, 4, and the denominator, 8, is 4.
Divide that numerator and denominator by 4.

$$\frac{4}{9} \times \frac{3}{8} = \frac{\overset{1}{\cancel{4}}}{9} \times \frac{3}{\underset{2}{\cancel{8}}}$$

Write the new fraction expression. The numerator 3 and the denominator 9 have a GCF of 3.
Divide that numerator and denominator by 3.

$$\frac{1}{\underset{3}{\cancel{9}}} \times \frac{\overset{1}{\cancel{3}}}{2} = \frac{1 \times 1}{3 \times 2}$$

Multiply the numerators and denominators.

$$\frac{1 \times 1}{3 \times 2} = \frac{1}{6}$$

Kristi put $\frac{1}{6}$ cup of yogurt in her smoothie.

EXERCISES

Find each product. Write your answer in simplest form.

1. $\frac{1}{3} \times \frac{4}{5}$

2. $\frac{3}{4} \times \frac{1}{2}$

3. $\frac{2}{3} \times \frac{1}{5}$

4. $\frac{4}{5} \times \frac{2}{5}$

5. $\frac{1}{4} \times \frac{4}{5}$

6. $\frac{5}{6} \times \frac{3}{5}$

7. $\frac{4}{7} \times \frac{3}{10}$

8. $\frac{3}{4} \times \frac{2}{3}$

9. $\frac{5}{6} \times \frac{7}{10}$

10. A cinnamon roll recipe requires $\frac{5}{8}$ cup of raisins. Beth wanted to make $\frac{1}{2}$ of the recipe of cinnamon rolls. She only needed $\frac{1}{2}$ of the raisins. How many cups of raisins does she need for half of the cinnamon roll recipe?

11. Four cheerleaders had mastered $\frac{1}{2}$ of their dance routine. However, Sadie was behind. She had only mastered $\frac{1}{2}$ of what the other four cheerleaders had mastered. What fraction of the dance routine had Sadie mastered?

Simplify the numbers in each fraction before multiplying. Find each product.

12. $\frac{1}{15} \times \frac{5}{6}$

13. $\frac{2}{3} \times \frac{3}{5}$

14. $\frac{4}{5} \times \frac{3}{8}$

15. $\frac{2}{7} \times \frac{3}{4}$

16. $\frac{1}{5} \times \frac{10}{13}$

17. $\frac{3}{4} \times \frac{12}{13}$

18. $\frac{4}{15} \times \frac{3}{8}$

19. $\frac{15}{16} \times \frac{4}{9}$

20. $\frac{6}{7} \times \frac{7}{12}$

21. Da-Shawn made $\frac{3}{4}$ of his shots at basketball practice. Trent made $\frac{2}{5}$ of the number of shots Da-Shawn made. What fraction of shots did Trent make?

22. Camden completed $\frac{12}{15}$ of his math homework in thirty minutes. Catira completed $\frac{3}{4}$ of what Camden had completed in the same amount of time. What fraction of math homework had Catira completed in thirty minutes?

23. Raynesha biked for $\frac{3}{4}$ hour yesterday. Today she plans to bike for $\frac{5}{6}$ of the amount of time she did yesterday. How long does she plan to bike today?

REVIEW

Find the GCF of each pair of numbers.

24. 32 and 48

25. 20 and 30

26. 44 and 33

Use the given measurements to find the perimeter of each polygon.

27.
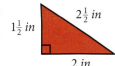
$1\frac{1}{2}$ in
$2\frac{1}{2}$ in
2 in

28.

SQUARE
$\frac{4}{5}$ in

29.
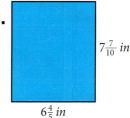
$7\frac{7}{10}$ in
$6\frac{4}{5}$ in

TIC-TAC-TOE ~ MULTIPLICATION OF MULTIPLE FRACTIONS

Follow the same procedure for multiplying two fractions when multiplying three or more fractions. Change any mixed numbers to improper fractions before multiplying.

Method 1: Multiply the numerators. Multiply the denominators. Simplify the fraction.

Example: $\frac{1}{2} \times \frac{2}{6} \times \frac{3}{4}$ $\qquad$ $\frac{1}{2} \times \frac{2}{6} \times \frac{3}{4} = \frac{1 \times 2 \times 3}{2 \times 6 \times 4} = \frac{6}{48} = \frac{1}{8}$

OR

Method 2: Simplify before multiplying. Divide one numerator and one denominator by a common factor. Repeat with other numerators and denominators as applicable. Multiply the numerators and denominators.

Example: $\frac{1}{2} \times \frac{2}{6} \times \frac{3}{4}$ $\qquad$ $\frac{1}{2} \times \frac{2}{6} \times \frac{3}{4} = \frac{1 \times 1 \times 1}{1 \times 2 \times 4} = \frac{1}{8}$

Write twenty different fractions or mixed numbers on small pieces of paper. Draw three or more fractions at random and write a multiplication problem with these fractions. Find the product. Repeat ten times using different combinations of fractions. Challenge yourself by selecting a set of four fractions at least twice.

TIC-TAC-TOE ~ CHOOSE YOUR OWN ADVENTURE

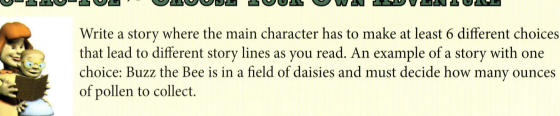

Write a story where the main character has to make at least 6 different choices that lead to different story lines as you read. An example of a story with one choice: Buzz the Bee is in a field of daisies and must decide how many ounces of pollen to collect.

At each point where the character must make a choice, put two different choices of expressions that use fractions. The reader of the story chooses which of the expressions to solve. If they solve one, the character makes one choice. If they solve the other, the character makes the other choice. Example: Buzz the Bee is in a field of daisies and must decide how many ounces of pollen to collect.

CHOICE 1: $\frac{5}{6} \times \frac{3}{10}$ $\qquad\qquad$ **CHOICE 2:** $\frac{5}{6} \div \frac{2}{3}$

If the reader chooses CHOICE 1, Buzz the Bee collects $\frac{1}{4}$ ounce of pollen. The reader is instructed to turn to page____ where that answer is used in the story. The answer causes something to happen and another choice has to be made.

If the reader chooses CHOICE 2, Buzz the Bee collects $1\frac{1}{4}$ ounces of pollen. The reader is instructed to turn to page ____ where that answer is used in the story. The answer causes something to happen and another choice has to be made.

DIVIDING FRACTIONS WITH MODELS

LESSON 5.3

🎯 Use models to divide fractions.

A group of students wanted to play a game of football in an open field. They marked off 100 yards for the entire football field. They needed to know how many 10-yard sections the 100 yard field covered. This can be written in three different ways.

$$10\overline{)100} \quad \text{or} \quad \frac{100}{10} \quad \text{or} \quad 100 \div 10$$

The students divided the 100 yards into 10-yard sections. It looked something like this diagram.

The students saw that 100 yards divided by 10 yards gave them 10 sections of 10 yards each.

$$\text{divisor} \rightarrow 10\overline{)100} \begin{array}{l} \leftarrow \text{quotient} \\ \leftarrow \text{dividend} \end{array} \qquad \frac{100}{10} = 10 \begin{array}{l} \leftarrow \text{dividend} \\ \leftarrow \text{quotient} \\ \leftarrow \text{divisor} \end{array} \qquad 100 \div 10 = 10 \begin{array}{l} \leftarrow \text{dividend} \\ \leftarrow \text{quotient} \\ \leftarrow \text{divisor} \end{array}$$

In each problem, 100 is the dividend, the number being divided. The divisor is 10, the number used to divide. The answer to the problem is called the quotient.

Dividing requires you to find how many groups of one number fit into another number. The same approach applies when dividing fractions.

EXAMPLE 1	**Maelynn needs to measure $\frac{3}{4}$ cup of milk, but only has a $\frac{1}{4}$ cup measuring cup. How many times will she need to fill it?**

SOLUTION

Write the problem.

$$\frac{3}{4} \div \frac{1}{4}$$

How many times does $\frac{1}{4}$ fit into $\frac{3}{4}$?

Draw a picture that represents the dividend, $\frac{3}{4}$.

Circle sets that are the size of the divisor, $\frac{1}{4}$.

Count how many sets of the divisor fit in the dividend to find the quotient. There are 3 circled sets of $\frac{1}{4}$ in $\frac{3}{4}$.

$$\frac{3}{4} \div \frac{1}{4} = 3$$

Maelynn will need to fill the $\frac{1}{4}$ cup measuring cup 3 times.

Example 1 shows how to find the quotient of two fractions with the same denominator. What happens if the fractions have unlike denominators? It is necessary, in this case, to rename the fractions so they have common denominators.

<table>
<tr><td>**EXAMPLE 2**</td><td>Clint needs to make a platform that is $\frac{3}{4}$ inch thick. He has boards that are each $\frac{3}{8}$ of an inch thick. How many boards does he need to make the platform?</td></tr>
</table>

SOLUTION

Write the problem.

$$\frac{3}{4} \div \frac{3}{8}$$

How many times does $\frac{3}{8}$ fit into $\frac{3}{4}$?

Rename one or both of the fractions so they have common denominators.

$$\frac{3}{4} = \frac{6}{8}$$

Draw a picture that represents the dividend, $\frac{3}{4}$ or $\frac{6}{8}$.

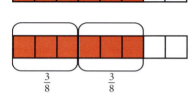

Circle sets that are the size of the divisor, $\frac{3}{8}$.

Count how many sets of the divisor fit in the dividend to find the quotient.
There are 2 circled sets of $\frac{3}{8}$ in $\frac{6}{8}$ or $\frac{3}{4}$.

$$\frac{3}{4} \div \frac{3}{8} = 2$$

Clint will need two boards, each $\frac{3}{8}$ inch thick, to make a $\frac{3}{4}$ inch thick platform.

EXPLORE!   **WHAT FITS?**

Dividend

Step 1: Choose one expression from the yellow box. (This example uses $\frac{1}{2} \div \frac{1}{8}$.)

Divisor

Step 2: If the two fractions in this expression have the same denominator, go to **Step 3.** If the two fractions have unlike denominators, find a common denominator.

$$\frac{1}{2} = \frac{4}{8}$$

Step 3: Draw a rectangle. Divide it into as many sections as the denominator of the dividend. If you renamed the dividend in **Step 2,** use the new fraction.

Step 4: Color in as many sections as the numerator of the dividend.

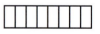

Step 5: Circle sets of your divisor in your drawing.

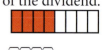

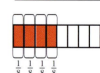

$\frac{1}{2} \div \frac{1}{4}$,	$\frac{1}{2} \div \frac{1}{8}$,	$\frac{2}{3} \div \frac{1}{6}$
$\frac{2}{3} \div \frac{2}{6}$,	$\frac{1}{4} \div \frac{1}{8}$,	$\frac{5}{8} \div \frac{1}{8}$
$\frac{3}{4} \div \frac{1}{8}$,	$\frac{1}{3} \div \frac{1}{6}$,	$\frac{2}{3} \div \frac{1}{3}$

Step 6: Count how many sets of the divisor are circled in your drawing.

Step 7: Write your drawing as an equation with an answer. $\frac{1}{2} \div \frac{1}{8} = \square$ ——— Quotient

Step 8: Use **Steps 1-7** to find the quotient of at least four different expressions from the yellow box.

EXERCISES

1. $\frac{6}{8} \div \frac{1}{4}$

   a. Rename one or both of the fractions so they have common denominators.
   b. Draw a rectangular model to represent the dividend.
   c. Circle sets in the rectangle that are the size of the divisor.
   d. How many sets of the divisor fit in the dividend? This is the quotient.

Use the procedure from Exercise 1 to find each quotient.

2. $\frac{4}{6} \div \frac{1}{6}$ **3.** $\frac{4}{7} \div \frac{2}{7}$ **4.** $\frac{9}{10} \div \frac{1}{10}$

5. $\frac{6}{10} \div \frac{1}{5}$ **6.** $\frac{6}{9} \div \frac{1}{3}$ **7.** $\frac{8}{12} \div \frac{2}{6}$

Write the equation to match each of the following models.

8.  _____ ÷ _____ = _____   **9.** _____ ÷ _____ = _____

10. _____ ÷ _____ = _____   **11.** _____ ÷ _____ = _____

12. _____ ÷ _____ = _____   **13.** _____ ÷ _____ = _____

14. Treva had a bottle that held $\frac{9}{16}$ ounces of liquid. She had a $\frac{3}{16}$ ounce measuring cup. How many measuring cups of liquid would she need to fill her bottle?

15. Taj took the monorail to his work which was $\frac{8}{9}$ mile away. The monorail stopped every $\frac{2}{9}$ mile. How many times did the train stop while Taj was on it?

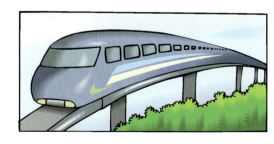

16. Jarrod built a tower of blocks that was $\frac{2}{3}$ meter tall. The blocks were each $\frac{1}{15}$ meter tall. How many blocks did he use?

17. Abir's mom sliced an orange. She gave Abir $\frac{2}{3}$ of the orange. Each slice was $\frac{1}{12}$ of the orange. How many slices of orange did Abir have?

REVIEW

Find each product. Write your answer in simplest form.

18. $\frac{1}{3} \times \frac{1}{2}$

19. $\frac{3}{5} \times \frac{5}{8}$

20. $\frac{2}{7} \times \frac{7}{8}$

Find the LCM of each set of numbers.

21. 4, 5

22. 3, 4, 6

23. 3, 5, 6

Find each difference using renaming.

24. $4\frac{3}{4} - 2\frac{7}{8}$

25. $5\frac{1}{8} - 3\frac{1}{3}$

26. $6\frac{3}{5} - 2\frac{11}{15}$

TIC-TAC-TOE ~ TEACH ME!

Make a flap book by folding a long sheet of paper lengthwise. Cut just to the fold to create two flaps (as shown in the diagram).

Label one flap "Multiplying Fractions Using Models." Label the other flap "Dividing Fractions Using Models."

Under the first flap, write directions to teach a classmate how to multiply fractions using models.

Under the second flap, write directions to teach a classmate how to divide fractions using models.

Have a parent or classmate try your directions to see if they work. Make changes to your directions so they are clear and useful, if needed.

DIVIDING FRACTIONS

Find quotients of expressions involving two fractions.

Drawing a model to find how many of one fraction fits into another fraction helps visualize the quotient. However, it becomes more difficult when the fractions do not divide evenly into each other. For example: $\frac{1}{2} \div \frac{3}{8}$.

If you were to draw this as in **Lesson 5.3**, your drawing would show that $\frac{3}{8}$ does not divide evenly into $\frac{1}{2}$.

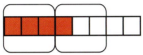

Reciprocals are used to divide fractions without models. Two numbers are reciprocals if their product is 1. To find the reciprocal of a fraction, "flip" the fraction. The numerator becomes the denominator and the denominator becomes the numerator.

$$\frac{3}{5} \xrightarrow{\text{Reciprocal}} \frac{5}{3} \qquad\qquad \frac{1}{4} \xrightarrow{\text{Reciprocal}} \frac{4}{1}$$

$$\checkmark \ \frac{3}{5} \times \frac{5}{3} = \frac{15}{15} = 1 \qquad\qquad \checkmark \ \frac{1}{4} \times \frac{4}{1} = \frac{4}{4} = 1$$

EXAMPLE 1

Find the reciprocals of the following numbers.

a. $\frac{2}{3}$ 　　　　　　b. $\frac{1}{7}$ 　　　　　　c. $\frac{5}{6}$

SOLUTIONS

a. Since $\frac{2}{3} \times \frac{3}{2} = \frac{6}{6} = 1$, the reciprocal of $\frac{2}{3}$ is $\frac{3}{2}$.

b. Since $\frac{1}{7} \times \frac{7}{1} = \frac{7}{7} = 1$, the reciprocal of $\frac{1}{7}$ is $\frac{7}{1}$.

c. Since $\frac{5}{6} \times \frac{6}{5} = \frac{30}{30} = 1$, the reciprocal of $\frac{5}{6}$ is $\frac{6}{5}$.

DIVIDING FRACTIONS

To divide by a fraction you must multiply by its reciprocal.

For any numbers *a*, *b*, *c* and *d*:
$$\frac{a}{b} \div \frac{c}{d} = \frac{a}{b} \times \frac{d}{c}$$

EXAMPLE 2

Find the value of $\frac{1}{2} \div \frac{2}{3}$.

SOLUTION

Find the reciprocal of the divisor.

$\frac{2}{3} \rightarrow \left(\frac{3}{2}\right)$ because $\frac{2}{3} \times \frac{3}{2} = 1$

Multiply the dividend by the reciprocal of the divisor.

$\frac{1}{2} \div \frac{2}{3} = \frac{1}{2} \times \frac{3}{2} = \frac{1 \times 3}{2 \times 2} = \frac{3}{4}$

$\frac{1}{2} \div \frac{2}{3} = \frac{3}{4}$

EXAMPLE 3

Yvette climbed $\frac{1}{2}$ the stairs at her house. Each stair was $\frac{1}{14}$ of the staircase. How many stairs had she climbed?

SOLUTION

Write the problem.

$\frac{1}{2} \div \frac{1}{14}$

Find the reciprocal of the divisor.

$\frac{1}{14} \rightarrow \left(\frac{14}{1}\right)$ because $\frac{1}{14} \times \frac{14}{1} = \frac{14}{14} = 1$

Multiply the dividend by the reciprocal of the divisor.

$\frac{1}{2} \div \frac{1}{14} = \frac{1}{2} \times \frac{14}{1} = \frac{1 \times 14}{2 \times 1} = \frac{14}{2} = 7$

☑ Check using a model.

Yvette had climbed 7 stairs.

EXAMPLE 4

At the beginning of this lesson you saw a drawing of the model for $\frac{1}{2} \div \frac{3}{8}$. Find this quotient.

SOLUTION

Find the reciprocal of the divisor.

$\frac{3}{8} \rightarrow \left(\frac{8}{3}\right)$ because $\frac{3}{8} \times \frac{8}{3} = \frac{24}{24} = 1$

Multiply the dividend by the divisor's reciprocal.

$\frac{1}{2} \div \frac{3}{8} = \frac{1}{2} \times \frac{8}{3} = \frac{1 \times 8}{2 \times 3} = \frac{8}{6}$

Simplify.

$\frac{8}{6} = \frac{4}{3} = 1\frac{1}{3}$

☑ Check using the model.

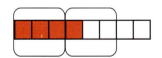

$\frac{1}{2} \div \frac{3}{8} = 1\frac{1}{3}$

Notice how the model shows the correct answer. There are $1\frac{1}{3}$ red sections circled.

EXERCISES

Find each quotient. Write your answer in simplest form.

1. $\frac{2}{3} \div \frac{1}{3}$

2. $\frac{6}{7} \div \frac{3}{7}$

3. $\frac{2}{5} \div \frac{1}{10}$

4. $\frac{1}{2} \div \frac{1}{16}$

5. $\frac{4}{5} \div \frac{2}{15}$

6. $\frac{7}{8} \div \frac{1}{8}$

7. $\frac{1}{2} \div \frac{1}{8}$

8. $\frac{3}{4} \div \frac{2}{8}$

9. $\frac{2}{3} \div \frac{2}{9}$

10. Shiloh drives $\frac{3}{4}$ mile to school each day. There is a stop sign every $\frac{3}{8}$ mile. How many times does Shiloh have to stop on her way to school?

11. When measuring ingredients:
   a. How many $\frac{1}{4}$ teaspoons make $\frac{1}{2}$ teaspoon?
   b. How many $\frac{1}{3}$ cups make $\frac{2}{3}$ cup?
   c. How many $\frac{1}{8}$ teaspoons make $\frac{1}{2}$ teaspoon?
   d. How many $\frac{1}{8}$ cups make $\frac{3}{4}$ cup?

Find each quotient. Write your answer in simplest form.

12. $\frac{4}{9} \div \frac{2}{3}$

13. $\frac{5}{7} \div \frac{1}{2}$

14. $\frac{5}{6} \div \frac{1}{5}$

15. $\frac{5}{9} \div \frac{1}{3}$

16. $\frac{4}{5} \div \frac{1}{3}$

17. $\frac{6}{7} \div \frac{3}{5}$

18. $\frac{1}{2} \div \frac{4}{5}$

19. $\frac{3}{4} \div \frac{1}{2}$

20. $\frac{2}{3} \div \frac{1}{2}$

21. Stefan has a board $\frac{2}{3}$ foot thick. He cuts $\frac{1}{8}$ foot thick sections out of it. How many sections will he cut?

22. Three-fourths of a cake remains after a birthday party. Each serving is $\frac{1}{16}$ of the cake. How many servings remain?

Write a division equation that matches each model.

23.  _____ ÷ _____ = _____

24. _____ ÷ _____ = _____

Find the value of each expression.

25. $\frac{3}{8} + \frac{1}{2}$

26. $\frac{7}{9} - \frac{1}{3}$

27. $4\frac{2}{5} + 1\frac{1}{3}$

28. $7\frac{1}{3} - 4\frac{1}{4}$

TIC-TAC-TOE ~ CELL PHONE PLANS

Step 1: Look up four different family plans for cell phones. Find the cost of each family plan (rounded to the nearest dollar) and the number of minutes allowed in each family plan.

Step 2: Figure out the fraction each person in your family represents.

Example: There are five people in your family. The total number of family members would be the denominator. Each person represents $\frac{1}{5}$.

Step 3: Multiply the fraction that represents one person by the total minutes in each cell phone plan. This is the number of minutes you would have if the total minutes were divided evenly among family members.

Step 4: Multiply the fraction that represents one person by the total price of each family plan (rounded to the nearest dollar) to find how much each plan costs per person.

Step 5: Create a poster to display your results. Choose the plan that you think is the best. Include your reasons for choosing that plan on the poster.

ESTIMATING PRODUCTS AND QUOTIENTS

LESSON 5.5

Estimate products and quotients using compatible numbers.

Finding exact answers to problems involving multiplying and dividing fractions takes time and requires the use of paper and pencil. You may not have these things available to you in many real-world situations. Compatible numbers are very important when estimating products and quotients. Compatible numbers are numbers that are easy to mentally compute.

EXPLORE!   4-H CLUB

The Barnyard Critters 4-H Club held a fundraiser for their club. Each member raised money separately. The club leader collected all of the money. The pie chart to the right shows the fraction of the money spent on each type of animal the club members raise.

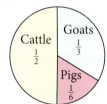

Barnyard Critters 4-H Club

Step 1: Parrish collected $28 during the fundraiser. He wants to determine the amount of his money that was spent on goats.
  a. What calculation could Parrish do to find the exact answer?
  b. Parrish wants to estimate the answer in his head. He changed the amount he collected to a number that is compatible with $\frac{1}{3}$. Which number, 27 or 29, is more compatible with $\frac{1}{3}$? Why?
  c. Use this number to determine the approximate amount of Parrish's funds spent on goats.

Step 2: Nichole raised $35. She wants to determine the amount of her money that will be spent on pigs.
  a. What calculation could Nichole do to find the exact answer?
  b. Nichole estimated the answer in her head. What whole number close to 35 is compatible with $\frac{1}{6}$? Why?
  c. Use this number to determine the approximate amount of Nichole's funds spent on pigs.

Step 3: The club members raised $149 overall. Approximate the amount of money spent on cows using a compatible number. What number did you choose and why?

Step 4: Use your answer from **Step 3.** Which statement would best express your answer and why?
  "The 4-H Club will spend a little less than $_____ on cows."
  "The 4-H Club will spend a little more than $_____ on cows."

Step 5: Estimate each of the following calculations using compatible numbers. Explain in words why you chose the numbers you did.

  a. $\frac{1}{3} \times 20$ **b.** $\frac{1}{5} \times 17$ **c.** $38 \times \frac{1}{8}$

Step 6: Give a real-world situation where compatible numbers can be used to multiply. Make up an example for that situation. Estimate the solution to your situation using compatible numbers.

EXAMPLE 1

Estimate the value of $\frac{1}{6} \times 20$.

SOLUTION

The number 20 cannot be equally divided into sixths.

List the multiples of 6.

Substitute the closest compatible number for 20 that is a multiple of 6. Find the value of $\frac{1}{6} \times 18$ by dividing 18 into 6 equal amounts.

$\frac{1}{6} \times 20 \approx 3$

18 is the closest multiple of 6

6: 6, 12, 18, 24, 30 …

$\frac{1}{6} \times 20$
↓
$\frac{1}{6} \times 18 = 3$

Estimate products or quotients with two mixed numbers by substituting whole compatible numbers for the mixed numbers.

EXAMPLE 2

Johanna has a $53\frac{1}{4}$ inch long piece of tubing for her science project. The project directions suggest cutting the tubing into $8\frac{2}{3}$ inch long pieces. About how many pieces will she be able to cut?

SOLUTION

Write the problem.

Round the divisor to the nearest whole number.

Change the dividend to the nearest multiple of the new divisor.

Johanna will be able to cut approximately 6 pieces of tubing.

$53\frac{1}{4} \div 8\frac{2}{3}$
↓
$53\frac{1}{4} \div 9$
↓
$54 \div 9 = 6$

EXAMPLE 3

Jack needed about $5\frac{3}{4}$ cups of dirt for each pot he was filling. He had $4\frac{1}{4}$ pots left to fill. Approximately how much dirt does he still need?

SOLUTION

Use compatible numbers to estimate.

Choose compatible numbers for each mixed number.

$$5\frac{3}{4} \times 4\frac{1}{4}$$
$$\downarrow \qquad \downarrow$$
$$6 \times 4$$

Solve the problem.

$6 \times 4 = 24$, so $5\frac{3}{4} \times 4\frac{1}{4} \approx 24$.

Jack still needs approximately 24 cups of dirt.

EXERCISES

1. Estimating is very useful in many situations.
   a. Describe one situation where you would use estimation rather than determining the exact answer.
   b. Describe one situation where you would NOT use estimation and only an exact answer would be appropriate.

2. Define "compatible number" in your own words.

Estimate each product using compatible numbers.

3. $\frac{1}{2} \times 23$

4. $\frac{1}{4} \times 15$

5. $\frac{2}{3} \times 19$

6. $\frac{1}{6} \times 47$

7. $\frac{3}{8} \times 17$

8. $\frac{2}{5} \times 11$

9. Aisha hit 46 pitches in batting practice. One-third of the hits were fly balls. About how many hits were fly balls?

10. Jeb bought 37 tickets for the carnival. He gave $\frac{1}{6}$ of the tickets to his brother. Approximately how many tickets did Jeb's brother get?

Estimate each product using compatible numbers.

11. $6\frac{1}{4} \times 3\frac{1}{8}$

12. $5\frac{1}{9} \times 5\frac{6}{7}$

13. $4\frac{3}{10} \times 6\frac{5}{6}$

14. $9\frac{1}{12} \times 2\frac{7}{9}$

15. $7\frac{2}{11} \times 2\frac{1}{5}$

16. $3\frac{13}{15} \times 2\frac{2}{9}$

Estimate each product using compatible numbers.

17. Mica followed a jam recipe that called for $5\frac{1}{4}$ cups of sugar. She wanted to make $5\frac{3}{4}$ batches of jam. About how much sugar would she need?

18. Joel was filling containers with cement. Each container held about $9\frac{1}{8}$ scoops of cement. He had $6\frac{8}{9}$ containers left to fill. Estimate how much cement he needs to fill the rest of the containers.

Estimate each quotient using compatible numbers.

19. $4\frac{1}{5} \div 1\frac{5}{6}$

20. $20\frac{7}{9} \div 2\frac{7}{8}$

21. $44\frac{3}{4} \div 5\frac{2}{17}$

22. $61\frac{1}{8} \div 10\frac{2}{15}$

23. $43\frac{1}{4} \div 6\frac{4}{5}$

24. $80\frac{1}{4} \div 9\frac{1}{7}$

25. Eric was a camp chef. He had $27\frac{7}{9}$ gallons of milk in the refrigerator. He used about $3\frac{5}{8}$ gallons per day. Approximately how many days will pass before he runs out of milk?

26. Zohar took $35\frac{3}{4}$ dollars from his bank account. He spent $4\frac{1}{5}$ dollars on Tuesday. He continues spending about the same amount each day. About how many days will it be until he needs to get more money?

REVIEW

Write each product in simplest form.

27. $\frac{1}{2} \times \frac{3}{4}$

28. $\frac{1}{3} \times \frac{3}{6}$

29. $\frac{2}{5} \times \frac{7}{8}$

Write each quotient in simplest form.

30. $\frac{7}{8} \div \frac{1}{2}$

31. $\frac{1}{2} \div \frac{1}{4}$

32. $\frac{5}{6} \div \frac{5}{8}$

TIC-TAC-TOE ~ FRACTION BINGO

Design a fraction BINGO game where players must correctly multiply or divide fractions.

Step 1: Create 65 fraction multiplication or division problems which have different answers.

Step 2: Use the 65 products or quotients from these problems to fill in at least ten BINGO cards with 25 spaces. Put a "free space" in the center space.

Step 3: Write a set of directions for your game. Bring the game to math class to play.

Find products or quotients of expressions that include fractions and whole numbers.

J'Marcus was up to bat nine times in his last two baseball games. One-third of his at-bats were hits. He wanted to determine how many hits he had during his nine at-bats. He must find the value of $\frac{1}{3} \times 9$ in order to find the answer. J'Marcus did this by drawing a model.

Each ball represents one of his nine at-bats.

He circled $\frac{1}{3}$ of the balls.

J'Marcus had hits in 3 of his 9 at-bats.

You can multiply or divide fractions and whole numbers using models. You can also follow the procedures you learned earlier in this Block to find the product or quotient. Before multiplying or dividing, you must write the whole number as a fraction.

A whole number can be written as a fraction by putting a 1 in the denominator. Two examples are shown below.

$$9 = \frac{9}{1} \qquad\qquad 15 = \frac{15}{1}$$

J'Marcus could have found how many hits he had in his last nine at bats by multiplying $\frac{1}{3} \times \frac{9}{1}$.

$$\frac{1}{3} \times \frac{9}{1} = \frac{1 \times 9}{3 \times 1} = \frac{9}{3} = 3$$

MULTIPLYING OR DIVIDING FRACTIONS AND WHOLE NUMBERS

1. Write the whole number as a fraction with a denominator of 1.
2. Multiply or divide.
3. Write the answer in simplest form.

EXAMPLE 1

Five-eighths of Michael's pitches in his last baseball game were strikes. If he threw 88 pitches, how many of these were strikes?

SOLUTION

Write the problem.

$$\frac{5}{8} \times 88$$

Write the whole number as a fraction.

$$\frac{5}{8} \times \frac{88}{1}$$

Multiply the fraction.

$$\frac{5}{8} \times \frac{88}{1} = \frac{5 \times \overset{11}{\cancel{88}}}{\underset{1}{\cancel{8}} \times 1} = \frac{55}{1} = 55$$

Fifty-five of Michael's pitches were strikes.

EXAMPLE 2

Find $7 \div \frac{3}{5}$.

SOLUTION

Write the whole number as a fraction.

$$\frac{7}{1} \div \frac{3}{5}$$

Divide.

$$\frac{7}{1} \div \frac{3}{5} = \frac{7}{1} \times \frac{5}{3} = \frac{7 \times 5}{1 \times 3} = \frac{35}{3} = 11\frac{2}{3}$$

$$7 \div \frac{3}{5} = 11\frac{2}{3}$$

EXAMPLE 3

Kaelani needed 3 cups of chocolate chips to make cookies. She only had a $\frac{1}{4}$ cup measure. How many $\frac{1}{4}$ cup measures would she need to make 3 cups?

SOLUTION

Write the problem.

$$3 \div \frac{1}{4}$$

Model.

Write the whole number as a fraction.

$$\frac{3}{1} \div \frac{1}{4}$$

Divide.

$$\frac{3}{1} \div \frac{1}{4} = \frac{3}{1} \times \frac{4}{1} = \frac{3 \times 4}{1 \times 1} = \frac{12}{1} = 12$$

Kaelani needs twelve $\frac{1}{4}$ cup measures of chocolate chips to equal 3 cups.

EXERCISES

Find each product.

1. $\frac{1}{4} \times 16$

2. $\frac{2}{3} \times 15$

3. $\frac{5}{6} \times 18$

4. $\frac{3}{7} \times 20$

5. $\frac{4}{9} \times 12$

6. $\frac{3}{4} \times 11$

7. The short track speed skating was dominated by South Korea in the 2006 winter Olympics at Torino, Italy. They took $\frac{5}{12}$ of the 24 medals awarded. How many medals did South Korea win?

8. Captain University's 2006-2007 men's basketball team won $\frac{2}{5}$ of their 30 games. How many games did they win?

9. Stacia washed windows at an office building. She had 24 windows to wash. At three o'clock Stacia had washed $\frac{4}{5}$ of the windows. How many windows had she washed so far?

10. A math teacher had 29 homework assignments to correct. She had corrected $\frac{3}{4}$ of the homework before she went home. How many assignments had she corrected?

Find each quotient.

11. $5 \div \frac{1}{2}$

12. $6 \div \frac{3}{4}$

13. $9 \div \frac{3}{5}$

14. $7 \div \frac{3}{4}$

15. $8 \div \frac{2}{3}$

16. $3 \div \frac{2}{7}$

17. Jessamyn ordered five pizzas. Jessamyn ordered enough so that each person at her party could eat $\frac{1}{3}$ of a pizza. How many people did she serve?

18. Jed bought six small bags of candy for his class. Each person received $\frac{1}{4}$ of a bag of candy. How many people got candy?

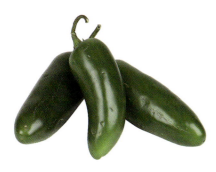

19. It took 15 gallons of gasoline to fill Reggie's gas tank. Reggie said his car used $\frac{2}{3}$ gallon of gas to drive to work and back. How many trips to work and back could Reggie make on one tank of gas?

20. Marita uses $\frac{3}{4}$ of a jalapeno pepper for each batch of fresh salsa. She has 13 jalapeno peppers. How many batches of salsa can she make?

Measure the sides of each polygon to the nearest sixteenth of an inch. Find each perimeter.

21.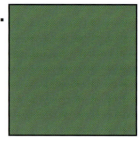

22.

23.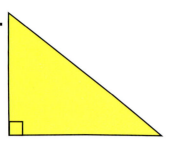

Estimate each product or quotient using compatible numbers.

24. $33 \div 8\frac{1}{8}$

25. $17 \div 3\frac{1}{3}$

26. $4\frac{2}{3} \times 8\frac{5}{6}$

27. $9\frac{4}{7} \div 2\frac{1}{8}$

28. $11\frac{1}{12} \times 3\frac{5}{7}$

29. $203 \div 19\frac{1}{9}$

TIC-TAC-TOE ~ HOW MANY MINUTES?

Karen spent $\frac{2}{3}$ hour practicing piano. She practiced a song called "Jazz Time" for $\frac{1}{2}$ of this time. How many minutes did she spend practicing "Jazz Time"?

Multiply the two fractions to figure out the fraction of an hour she spent practicing "Jazz Time."

Example: $\frac{1}{2} \times \frac{2}{3} = \frac{2}{6} = \frac{1}{3}$ Karen spent $\frac{1}{3}$ hour practicing "Jazz Time."

In order to figure out how many minutes Karen spent practicing this song, multiply the fraction of an hour by the number of minutes in an hour (60).

Example: $\frac{1}{3} \times 60 = 20$ Karen spent 20 minutes practicing "Jazz Time."

How many minutes are in…

1. $\frac{1}{4}$ hour?

2. $\frac{1}{6}$ hour?

3. $\frac{1}{10}$ hour?

4. $\frac{3}{4}$ hour?

5. $\frac{2}{3}$ hour?

6. $\frac{7}{12}$ hour?

7. $\frac{1}{3}$ of $\frac{1}{2}$ hour?

8. $\frac{5}{6}$ of $\frac{1}{2}$ hour?

9. $\frac{1}{3}$ of $\frac{1}{4}$ hour?

10. $\frac{1}{2}$ of $\frac{1}{6}$ hour?

11. $\frac{3}{8}$ of $\frac{2}{3}$ hour?

12. $\frac{7}{9}$ of $\frac{3}{4}$ hour?

MULTIPLYING AND DIVIDING MIXED NUMBERS

LESSON 5.7

 Find products and quotients of expressions that include mixed numbers.

You have learned how to find the products and quotients of two fractions or a fraction and a whole number in this Block. You will learn how to find products and quotients of expressions involving mixed numbers in this lesson.

EXPLORE!   **SCRABOOKING**

Katelyn enjoys putting her photos into scrapbooks. When she works diligently, she completes $3\frac{1}{2}$ pages each hour.

Step 1: Katelyn's relatives are coming in $2\frac{1}{2}$ hours. Katelyn wants to know how many pages she can finish before they arrive.
  a. Write a math problem that will help her determine this.
  b. Change each mixed number to an improper fraction and calculate.
  c. Simplify the answer. Write it in a complete sentence.

Step 2: Katelyn plans to complete 14 pages of her scrapbook tomorrow. How many hours will it take her?
  a. Write the problem.
  b. Change both the whole number and mixed number to fractions and calculate.
  c. Simplify the answer. Write the answer in a complete sentence.

Step 3: Katelyn spent a total of $9\frac{1}{3}$ hours on her scrapbook last week. How many pages did she complete?

Step 4: Stephen is just learning to scrapbook. He can finish $2\frac{3}{4}$ pages each hour. He worked with Katelyn for the entire $9\frac{1}{3}$ hours last week. How many pages did he complete?

Step 5: Describe in words the steps to take when multiplying or dividing two mixed numbers.

MULTIPLYING AND DIVIDING MIXED NUMBERS

1. Write each mixed or whole number as a fraction.
2. Multiply or divide using the improper fractions.
3. Write the answer in simplest form.

EXAMPLE 1

Find the value of $7\frac{1}{3} \times 1\frac{1}{4}$.

SOLUTION

Change each mixed number to an improper fraction.

$$7\frac{1}{3} = \frac{22}{3} \qquad 1\frac{1}{4} = \frac{5}{4}$$

Rewrite and multiply.

$$\frac{\overset{11}{\cancel{22}}}{3} \times \frac{5}{\underset{2}{\cancel{4}}} = \frac{55}{6}$$

Write as a mixed number.

$$\frac{55}{6} = 9\frac{1}{6}$$

$$7\frac{1}{3} \times 1\frac{1}{4} = 9\frac{1}{6}$$

EXAMPLE 2

Brennan bought $7\frac{3}{8}$ pounds of salmon for his family. Each portion was about $\frac{1}{2}$ pound of salmon. How many portions could he serve?

SOLUTION

Write the problem.

$$7\frac{3}{8} \div \frac{1}{2}$$

Write the mixed number as an improper fraction.

$$7\frac{3}{8} = \frac{59}{8}$$

Divide.

$$\frac{59}{8} \div \frac{1}{2} = \frac{59}{8} \times \frac{2}{1} = \frac{59 \times 2}{8 \times 1} = \frac{118}{8} = 14\frac{6}{8}$$

Simplify.

$$14\frac{6}{8} = 14\frac{3}{4}$$

Brennan could serve $14\frac{3}{4}$ portions of salmon.

To find the reciprocal of a whole number, write the whole number as a fraction over 1. Then "flip" the fraction.

$$3 = \frac{3}{1} \xrightarrow{\text{Reciprocal}} \frac{1}{3}$$

EXAMPLE 3

Isaac built a tree house for his daughter. He cut a board into 5 equal pieces. How long is each piece of board if the original board was $11\frac{1}{2}$ feet long?

SOLUTION

Write the problem.

$$11\frac{1}{2} \div 5$$

Write each whole and mixed number as an improper fraction.

$$11\frac{1}{2} = \frac{23}{2} \qquad 5 = \frac{5}{1}$$

Multiply by the reciprocal of the divisor.

$$\frac{23}{2} \div \frac{5}{1} = \frac{23}{2} \times \frac{1}{5} = \frac{23}{10}$$

Change into a mixed number.

$$\frac{23}{10} = 2\frac{3}{10}$$

Each equal piece of the board is $2\frac{3}{10}$ feet long.

EXERCISES

Find each product. Write each answer in simplest form.

1. $2\frac{1}{8} \times 2$

2. $3 \times 2\frac{2}{9}$

3. $2\frac{1}{5} \times 5$

4. $4\frac{2}{3} \times \frac{4}{5}$

5. $\frac{1}{3} \times 3\frac{3}{8}$

6. $\frac{5}{6} \times 2\frac{1}{3}$

7. $2\frac{2}{5} \times 1\frac{2}{3}$

8. $1\frac{1}{6} \times 3\frac{1}{2}$

9. $5\frac{1}{2} \times 1\frac{2}{3}$

10. Naomi used $4\frac{3}{4}$ tablespoons of ground coffee for each pot of coffee she made. She made one pot of coffee each day, Monday through Friday. How many tablespoons of ground coffee did she use altogether?

11. Yani made cookies. He used $2\frac{1}{4}$ cups of flour per batch of cookies. He made $3\frac{1}{2}$ batches of cookies. How much flour did he use?

Find each quotient. Write each answer in simplest form.

12. $6\frac{2}{5} \div 2$

13. $5 \div 1\frac{1}{4}$

14. $9\frac{2}{5} \div 3$

15. $3\frac{1}{9} \div \frac{1}{3}$

16. $2\frac{5}{6} \div \frac{3}{8}$

17. $4\frac{1}{6} \div \frac{3}{4}$

18. $4\frac{8}{9} \div 1\frac{1}{3}$

19. $3\frac{1}{2} \div 1\frac{1}{4}$

20. $5\frac{1}{3} \div 1\frac{2}{3}$

21. Ty edged a flower bed with paving stones. Each paving stone was $5\frac{3}{4}$ inches long. The length of the flower bed was $40\frac{1}{4}$ inches long. How many paving stones did Ty need?

22. J.D. built a tower out of blocks with his nephew. Each block was $1\frac{3}{4}$ inches tall. They built a $43\frac{3}{4}$ inch tall tower. How many blocks did they use?

REVIEW

Find the value of each expression.

23. 23.53×9.8

24. $103.7 \div 12.2$

25. $749.48 \div 16.4$

26. $3 \times \frac{2}{3}$

27. $\frac{14}{15} \div \frac{1}{5}$

28. $9 \times \frac{4}{5}$

29. $8 \div \frac{3}{4}$

30. $\frac{2}{3} \times \frac{9}{10}$

31. $11 \div \frac{2}{3}$

TIC-TAC-TOE ~ ADVERTISEMENTS

Step 1: Look online or in a newspaper for "Buy One, Get One Half Off" advertisements. Find at least five different advertisements. Print them or cut them out.

Step 2: Choose an item from one advertisement.

Step 3: Record the price of the first item (full price). Round to the nearest dollar.

Step 4: Record the price of the second identical item (half-off price). Round to the nearest dollar.

Step 5: Find the price per item by averaging the two prices $\frac{\text{full price + half price}}{2}$.

Step 6: Repeat **Steps 2-5** for four more items from different advertisements.

Step 7: Use the advertisements and your calculations to make a collage. Display the advertisements, the price of one item, the price of the second item at half-off, and the average price per item.

TIC-TAC-TOE ~ CHANGING RECIPES

Step 1: Choose a favorite recipe with at least five ingredients.

Step 2: Find the amounts of ingredients needed to cut the recipe in half. Rewrite the recipe using your results.

Step 3: Figure out the ingredient amounts you would need for $1\frac{1}{2}$ times the recipe. Rewrite the recipe using your results.

Step 4: Rewrite the recipe tripling the ingredients.

TIC-TAC-TOE ~ LEARNING WITH LYRICS

Write a song that teaches about common errors that might be made when adding, subtracting, multiplying or dividing fractions.

Put the song to music (it can be sung to a familiar tune or original music).

Sing it for your teacher or record it.

Vocabulary

compatible numbers

reciprocal

Use models to multiply fractions.
Find products of expressions involving two fractions.
Use models to divide fractions.
Find quotients of expressions involving two fractions.
Estimate products and quotients using compatible numbers.
Find products or quotients of expressions that include fractions and whole numbers.
Find products and quotients of expressions that include mixed numbers.

Lesson 5.1 ~ Multiplying Fractions with Models

Write an equation to match each model.

1.

2.

3.

Draw a model to represent each expression. Write the answer to each expression in simplest form.

4. $\frac{1}{4} \times \frac{1}{2}$

5. $\frac{1}{2} \times \frac{2}{5}$

6. $\frac{2}{3} \times \frac{1}{4}$

Lesson 5.2 ~ Multiplying Fractions

Find each product. Write the answer in simplest form.

7. $\frac{2}{3} \times \frac{3}{8}$

8. $\frac{7}{8} \times \frac{1}{4}$

9. $\frac{5}{6} \times \frac{2}{3}$

10. $\frac{4}{5} \times \frac{1}{8}$

11. $\frac{1}{2} \times \frac{3}{5}$

12. $\frac{5}{9} \times \frac{1}{5}$

13. Izaya washed $\frac{1}{3}$ of the cars at the car wash. Todd washed $\frac{3}{8}$ as many as Izaya washed. What fraction of the total cars did Todd wash?

14. Cindy bought $\frac{1}{3}$ yard of fabric. She only needs $\frac{2}{3}$ of this fabric for the project she's making. What fraction of a yard will she use?

15. Nadine ate $\frac{1}{4}$ of a chicken basket. Suzanne ate $\frac{1}{3}$ of the amount that Nadine ate. What fraction of the chicken basket did Suzanne eat?

Lesson 5.3 ~ Dividing Fractions with Models

Write an equation to match each model.

16.

17.

18.

Draw a model to represent each expression. Write the answer to the equation.

19. $\frac{3}{8} \div \frac{1}{8}$

20. $\frac{4}{5} \div \frac{2}{5}$

21. $\frac{1}{3} \div \frac{1}{6}$

Lesson 5.4 ~ Dividing Fractions

Find each quotient. Write your answer in simplest form.

22. $\frac{4}{5} \div \frac{1}{5}$

23. $\frac{1}{2} \div \frac{1}{8}$

24. $\frac{4}{5} \div \frac{1}{10}$

25. $\frac{3}{7} \div \frac{1}{2}$

26. $\frac{5}{6} \div \frac{1}{8}$

27. $\frac{3}{5} \div \frac{2}{3}$

28. Jordyn has $\frac{7}{8}$ liter of water. She has bottles that each hold $\frac{1}{2}$ liter of water. How many bottles can she fill?

29. Evie drives $\frac{2}{3}$ mile to work each day. She sees students waiting at bus stops every $\frac{1}{6}$ mile. How many times does she pass students waiting at bus stops on one trip to work?

30. Joseph put $\frac{3}{4}$ gallon of marinara sauce into jars. Each jar holds $\frac{1}{8}$ gallon. How many jars of marinara sauce does Joseph have?

Lesson 5.5 ~ Estimating Products and Quotients

Estimate each product or quotient.

31. $\frac{1}{2} \times 21$

32. $28\frac{1}{6} \div 4\frac{1}{3}$

33. $\frac{1}{8} \times 17$

34. $16\frac{1}{9} \div 5\frac{2}{11}$

35. $71\frac{4}{9} \div 8\frac{2}{9}$

36. $5\frac{1}{10} \times 3\frac{1}{3}$

37. Tamira entered a relay race where the team must swim a total of $35\frac{1}{8}$ laps. The four members of the team each swam equal distances in the race. About how many laps did each relay team member swim?

38. Nathaniel had 17 carrot sticks. He gave $\frac{1}{4}$ of the carrot sticks to his brother. Approximately how many carrot sticks did his brother get?

Lesson 5.6 ~ Multiplying and Dividing Fractions and Whole Numbers

Find each product or quotient. Write in simplest form.

39. $\frac{1}{8} \times 64$

40. $\frac{2}{5} \times 20$

41. $9 \div \frac{3}{4}$

42. $15 \div \frac{3}{5}$

43. $\frac{3}{4} \times 34$

44. $32 \div \frac{2}{3}$

45. $25 \div \frac{5}{8}$

46. $\frac{1}{5} \times 22$

47. $12 \times \frac{2}{5}$

48. Mrs. Jenkins tore off 7 pieces of butcher paper to decorate her bulletin boards. Each piece of paper was $\frac{3}{4}$ of a yard long. What was the total length of paper she tore off?

Lesson 5.7 ~ Multiplying and Dividing Mixed Numbers

Find each product or quotient. Write in simplest form.

49. $5\frac{1}{8} \times \frac{2}{3}$

50. $5\frac{3}{5} \div 1\frac{1}{2}$

51. $5\frac{1}{4} \div 2\frac{1}{2}$

52. $2\frac{1}{8} \div \frac{1}{3}$

53. $3\frac{1}{3} \times 1\frac{2}{3}$

54. $4\frac{2}{3} \times 2\frac{3}{4}$

55. Lani is given $19\frac{1}{2}$ dollars for lunches. She spends $3\frac{1}{4}$ dollars each day. How many days will it be until she runs out of money?

56. Sakaiya uses $4\frac{1}{4}$ cups of sugar for every batch of freezer jam. He makes $1\frac{1}{2}$ batches of jam. How many cups of sugar does he use?

CAREER FOCUS

JULIE
REPORTER

I am a news reporter. I talk to people to find the latest or most interesting news in my city. Then I write stories for the newspaper and the newspaper website. I also record or videotape people in the news to broadcast on the website. It is a fast-paced job. No two days are ever the same. I also know my city and state very well because reporters are always on the move looking for news.

I use math, including addition, subtraction, percentages and averages, almost every day. Numbers help me understand changes in our city, such as how many new people moved here from another state, or how many people voted for one candidate. Reporters study financial reports to understand if the businesses in our city are doing well or poorly, or if there is enough money to pay for more schools. Numbers don't lie.

I received a Bachelor of Arts degree in journalism to become a journalist. I also studied history. Some reporters study English, geology, political science or law. What we share is a common desire to take complicated information and present it in a way that people can understand. Reporters can also work for television and radio stations. Every city needs reporters. Reporters in small towns may earn $20,000 a year while those in larger cities earn $45,000 to $85,000 per year.

Being a reporter is like holding a magic key to enter any door or world you wish. I can go into a hospital operating room to write a story about a doctor, or ride in an experimental airplane. I can meet famous people. But the most important thing I do is find hidden news that, once it becomes known, makes the world a better place.

CORE FOCUS ON DECIMALS & FRACTIONS
BLOCK 6 ~ AREA AND VOLUME

LESSON 6.1   AREA WITH FRACTIONS --- 169
              EXPLORE! TRIANGLE AREA

LESSON 6.2   AREA AND PERIMETER WITH DECIMALS -------------------------------- 174

LESSON 6.3   AREAS OF COMPOSITE FIGURES -------------------------------------- 178

LESSON 6.4   NETS AND SURFACE AREAS --------------------------------------- 182
              EXPLORE! NETTING A SOLID

LESSON 6.5   VOLUME WITH FRACTIONAL DIMENSIONS ------------------------- 189
              EXPLORE! MEASURING VOLUME

REVIEW   BLOCK 6 ~ AREA AND VOLUME ----------------------------------- 194

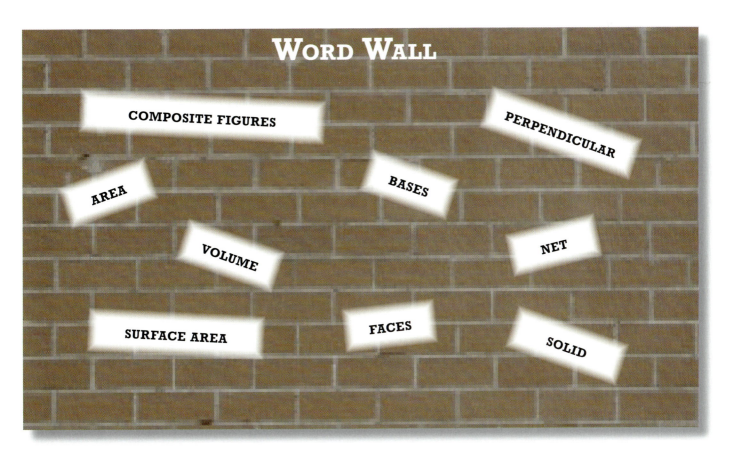

BLOCK 6 ~ AREA AND VOLUME
TIC-TAC-TOE

WHERE IN THE WORLD?

Create a poster showing 3-dimensional solids from the real world. Create nets for each real-world solid.

See page 188 for details.

NET PATTERNS

Make 2-dimensional net patterns of 3-dimensional shapes.

See page 193 for details.

COMPOSITE FIGURES

Design composite figures. Add measurements that include fractions and decimals.

See page 193 for details.

DREAM HOUSE

Draw a one story dream house. Find the square footage of the blueprint.

See page 173 for details.

VOCABULARY MEMORY

Create a vocabulary matching game to help students remember all the vocabulary words.

See page 188 for details.

SIMILARITY SWAP

Design boxes with the same surface areas but different volumes. Design boxes with the same volume but different surface area.

See page 188 for details.

BOX BLUES

Figure out the surface areas and volumes of multiple boxes of cereal for a company. Calculate costs for the company.

See page 198 for details.

LETTER TO THE EDITOR

Should the USA use metric measurement or customary measurement? Write a letter to the editor with reasons to back up your opinion.

See page 177 for details.

BEDROOM AREA

Find the area of your room and furniture in square meters. See if there are different ways to arrange the furniture.

See page 173 for details.

AREA WITH FRACTIONS

Calculate areas of rectangles, squares and triangles with lengths of sides that are fractions or mixed numbers.

A builder needs to know how many square feet of siding to buy for one side of a house. A carpet layer needs to know the square footage of a room so carpet can be ordered. An architect finds the square footage on blueprints in order to give final dimensions to the owner of a company wishing to build a new office building.

Area is the number of square units used to cover a surface. A rectangle with an area of 6 square units can also be written as 6 units².

2 units | 6 units² | Six square units cover the surface of the rectangle.
3 units

How do the length and width affect the area? One side is 2 units. The other side is 3 units. The area is 6 square units. The area of different shapes can be found using multiplication.

$$2 \text{ units} \times 3 \text{ units} = 6 \text{ square units}$$

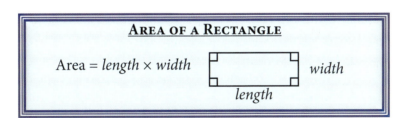

AREA OF A RECTANGLE

$$\text{Area} = length \times width$$

width

length

| EXAMPLE 1 | **Find the area of the rectangle.** |  $2\frac{1}{2}$ |

4

| SOLUTION | $\text{Area} = length \times width$ | $\text{Area} = 4 \times 2\frac{1}{2}$ |

Change the mixed number and whole number into improper fractions.

$4 = \frac{4}{1}$ and $2\frac{1}{2} = \frac{5}{2}$

Multiply.

$\text{Area} = \frac{4}{1} \times \frac{5}{2} = \frac{20}{2}$ or $\frac{\overset{2}{\cancel{4}}}{1} \times \frac{5}{\underset{1}{\cancel{2}}} = \frac{10}{1}$

Simplify.

$\frac{20}{2} = 10$ or $\frac{10}{1} = 10$

The area of the rectangle is 10 square units or 10 units².

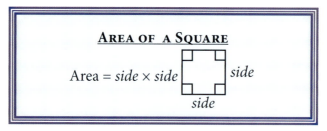

AREA OF A SQUARE

Area = *side* × *side*   *side*

side

EXAMPLE 2

Find the area of the square.

$1\frac{1}{2} ft$

SOLUTION

Area = *side* × *side*	Area $= 1\frac{1}{2} \times 1\frac{1}{2}$
Change the mixed number to an improper fraction.	$1\frac{1}{2} = \frac{3}{2}$
Multiply.	Area $= \frac{3}{2} \times \frac{3}{2} = \frac{9}{4}$
Simplify.	$\frac{9}{4} = 2\frac{1}{4}$

The area of the square is $2\frac{1}{4}$ square feet or $2\frac{1}{4} ft^2$.

EXPLORE!   **TRIANGLE AREA**

Step 1: Draw a rectangle on grid paper.

Step 2: Count the number of grid squares that make up the rectangle's area. Write this to the side of the rectangle as Area = ___ square units.

Step 3: Record the rectangle's length and width. Label these on the outside of the rectangle.

Step 4: Draw a diagonal line from one corner of the rectangle to the opposite corner to make a right triangle.

Step 5: What do you think is the area of each triangle formed? Explain your reasoning.

The length of the base and the height are used when finding the area of a triangle. The height of a triangle is a **perpendicular** line segment drawn from the base of the triangle to the opposite vertex. The base and height are perpendicular if they form a right angle where the two line segments meet.

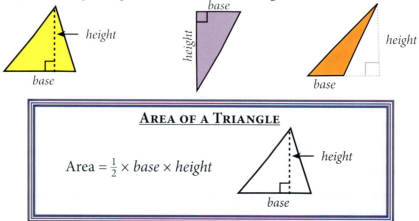

AREA OF A TRIANGLE

Area $= \frac{1}{2} \times$ *base* × *height*

height

base

EXAMPLE 3

Find the area of the triangle.

3 in

5 in

SOLUTION

Area = $\frac{1}{2}$ × base × height

Change the whole numbers into fractions.

Multiply.

Simplify.

Area = $\frac{1}{2}$ × 5 × 3

$5 = \frac{5}{1}$ and $3 = \frac{3}{1}$

Area = $\frac{1}{2} \times \frac{5}{1} \times \frac{3}{1} = \frac{15}{2}$

Area = $\frac{15}{2} = 7\frac{1}{2}$

The area of the triangle is $7\frac{1}{2}$ square inches or $7\frac{1}{2}$ in^2.

EXERCISES

Use the given measurements to find the area of each rectangle.

1.

2 ft

5 ft

2.

$3\frac{1}{2}$ units

3 units

3.

$2\frac{1}{4}$ units

$3\frac{3}{4}$ units

4.

$\frac{7}{8}$ in

$\frac{5}{8}$ in

5.

$\frac{1}{4}$ in

$1\frac{3}{8}$ in

6.

$1\frac{1}{2}$ in

$\frac{3}{4}$ in

 7. A carpet layer measured the floor in a room to be carpeted. The room was $20\frac{1}{3}$ ft long and $18\frac{3}{4}$ ft wide. What is the area of the floor in this room?

Use the given measurement to find the area of each square.

8.

4 units

9.

$2\frac{1}{3}$ units

10.

$6\frac{1}{2}$ in

11.

$3\frac{3}{4}$ in

12.

$1\frac{1}{8}$ in

13. Estaban has an end table with a square top. One side of the top of the table is $3\frac{1}{4}$ *ft* long. What is the area of the top of Estaban's table?

Find the area of the triangles using the given heights and bases.

14.

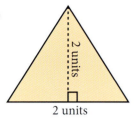

2 units

2 units

15.

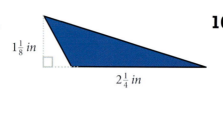

$1\frac{1}{8}$ *in*

$2\frac{1}{4}$ *in*

16.

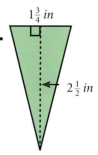

$1\frac{3}{4}$ *in*

$2\frac{1}{2}$ *in*

Measure the sides of the following polygons with a customary ruler to the nearest quarter inch. Find the area of each polygon.

17.

18.

19.

20.

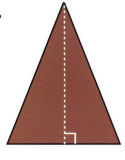

21.

22.

23. Two rectangles each have an area of 12 inches. Find the dimensions of two different rectangles that fit this description.

24. Two triangles each have an area of 12 inches. Find the dimensions of two triangles that fit this description.

REVIEW

Find the value of each expression.

25. $3\frac{1}{3} \times \frac{5}{6}$

26. $2\frac{1}{9} \times 2\frac{1}{2}$

27. $5\frac{1}{2} \times 2$

28. $6\frac{1}{5} \div \frac{2}{3}$

29. $2\frac{5}{7} \div \frac{5}{14}$

30. $5\frac{5}{8} \div 1\frac{1}{4}$

TIC-TAC-TOE ~ BEDROOM AREA

Step 1: Measure the width and length of your bedroom using a metric ruler or measuring tape, to the nearest centimeter. Make a scale drawing of your bedroom where a 1 cm line equals 1 meter. Write the measurements as decimals on the outside of the drawing of your bedroom.

Examples: length of 5 meters and 21 centimeters = 5.21 *m*
length of 5 meters and 3 centimeters = 5.03 *m*

Step 2: Measure, to the nearest centimeter, all large items that take up floor space in your room (bed, dresser, chair, desk, etc.) with a metric ruler or measuring tape. Measure at least three items. Draw these items on your scale drawing. Label them with their measurements.

Step 3: Calculate the area of your bedroom.

Step 4: Calculate the floor area each item covers in your bedroom. How much total floor space is left in your room?

Step 5: Rearrange your furniture on paper to maximize your floor space. Are there different ways you could arrange your furniture to give yourself more space? Draw different possible arrangements. You should consult parents or guardians before actually rearranging your bedroom.

TIC-TAC-TOE ~ DREAM HOUSE

Step 1: Use grid paper to draw a one-story dream house where the length of each side of a square equals $2\frac{1}{2}$ feet in real life. You must have at least 6 rooms. Make sure at least three rooms have an odd number of squares on the length and/or width.

Example:

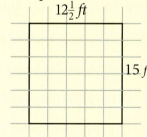

$12\frac{1}{2}\,ft$

15 *ft*   Game room = $12\frac{1}{2}\,ft \times 15\,ft$

$$\frac{25}{2} \times \frac{15}{1} = \frac{375}{2} = 187\frac{1}{2}\,ft^2$$

Step 2: Figure out the square footage of each room.

Step 3: Calculate the square footage of the entire house.

AREA AND PERIMETER WITH DECIMALS

LESSON 6.2

🎯 Calculate the perimeter and area of squares, rectangles and triangles using the metric system.

Metric measurements are most often written as decimals. Finding perimeter and area of figures that have metric measurements involves adding and multiplying decimals.

Celine's family landscaped their new backyard. The yard is a rectangle 50.5 meters long and 28.2 meters wide. They put a fence around the entire perimeter of the yard. How much fencing did they need?

You must find the perimeter to find the distance around the yard. Remember that opposite sides of rectangles are equal in length.

Perimeter = length + width + length + width
Perimeter = 50.5 + 28.2 + 50.5 + 28.2

$$\begin{array}{r} 50.5 \\ 28.2 \\ 50.5 \\ + 28.2 \\ \hline 157.4 \text{ meters} \end{array}$$

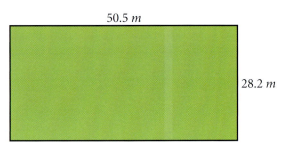

50.5 *m*

28.2 *m*

Celine's family also laid sod in the entire backyard. How many square meters of sod did they purchase?

To find out the amount of sod that will cover the backyard, you must find the area. The area of a rectangle is determined by multiplying the length times the width.

Area of a Rectangle = length × width
Area of Celine's Yard = 50.5 × 28.2

$$\begin{array}{r} 50.5 \\ \times\ 28.2 \\ \hline 1010 \\ 40400 \\ + 101000 \\ \hline 1424.10 \text{ square meters} \end{array}$$

Celine's family needed 157.4 meters of fencing and 1,424.1 square meters of sod for their backyard.

EXAMPLE 1

Find the area of the square using centimeters.

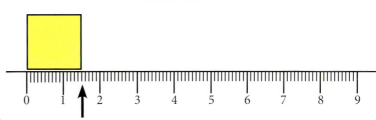

SOLUTION

Measure the sides of the shape.

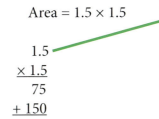

It is a square so each side measures 1.5 centimeters.

Square: Area = *side* × *side*   Area = 1.5 × 1.5

Multiply.

$$\begin{array}{r} 1.5 \\ \times\ 1.5 \\ \hline 75 \\ +\ 150 \\ \hline 2.25 \end{array}$$

> There is one place after the decimal point in each of the factors for a total of 2 places in the solution.

The area of the square is 2.25 square centimeters or 2.25 cm^2.

EXAMPLE 2

Use the given measurements to find the area of the triangle.

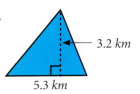

3.2 *km*

5.3 *km*

SOLUTION

The triangle area formula can be written two different ways.
$\frac{1}{2}$ × *base* × *height* or 0.5 × *base* × *height*
If the measurements are in decimals, use the formula that contains a decimal.

Area = 0.5 × *base* × *height*   Area = 0.5 × 5.3 × 3.2

Multiply the first two decimals.

$$\begin{array}{r} 0.5 \\ \times\ 5.3 \\ \hline 15 \\ +\ 250 \\ \hline 265 \rightarrow 2.65 \end{array}$$

> There is one digit after the decimal point in each of the factors for a total of two digits beyond the decimal point in the solution.

Multiply the answer from the first equation by the third number.

$$\begin{array}{r} 2.65 \\ \times\ 3.2 \\ \hline 530 \\ +\ 7950 \\ \hline 8480 \rightarrow 8.48 \end{array}$$

The area of the triangle is 8.48 square kilometers or 8.48 km^2.

EXERCISES

Use the given measurements to find the perimeter and area of each figure.

1.
1.4 *cm*
5.5 *cm*

2.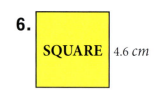
SQUARE   6.1 *cm*

3.
5.75 *m*
5.9 *m*   5.9 *m*
2.6 *m*

4.
6.9 km
4.9 km   3.2 km
4.9 km

5.
2.3 *mm*
6.5 *mm*

6.
SQUARE   4.6 *cm*

7. Use the block at the left.
   a. Find the perimeter.
   b. Find the area in centimeters.

8. Use the domino at the right.
   a. Find the perimeter.
   b. Find the area in centimeters.

Measure each side of each polygon to the nearest tenth of a centimeter. Find the perimeter of each polygon.

9.

10.

11.

12. The door to Karina's bedroom is 1.07 meters wide by 2.4 meters tall.
   a. What is the perimeter of Karina's door?
   b. What is the area of the door?

13. Omar walked 23.8 meters along one side of a square field.
   a. If he walked the perimeter of this field, how far would he travel?
   b. What is the area of the field?

14. A rectangular card is 14 *cm* long and 10.7 *cm* wide.
   a. What is the perimeter of this card?
   b. What is the area of this card?

Find the area of each polygon. Measure all necessary lengths to the nearest tenth of a centimeter.

15.

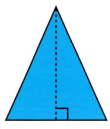

16.

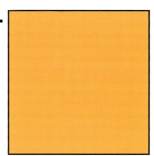

17.

REVIEW

Find the value of each expression. Write in simplest form.

18. $\frac{1}{3} \times \frac{3}{4}$

19. $\frac{7}{10} \div \frac{1}{5}$

20. $\frac{2}{3} + \frac{1}{8}$

21. $\frac{14}{15} - \frac{3}{5}$

22. $1\frac{2}{5} \div 2\frac{1}{10}$

23. $3\frac{1}{2} \times 1\frac{1}{2}$

24. $2\frac{1}{8} - 1\frac{3}{4}$

25. $4 \div \frac{1}{3}$

26. $2\frac{1}{4} \times 5$

27. A bag containing 15 pounds of grain is divided into $1\frac{2}{3}$ pound portions. How many portions are there?

28. A cat had 6 kittens in one litter. Each kitten weighed approximately $\frac{3}{10}$ of a pound. About how much did all six kittens weigh together?

AREAS OF COMPOSITE FIGURES

 Find the areas of composite figures by breaking them down into known shapes.

In previous lessons, you have learned how to find areas of three basic geometric shapes. The area of a shape is found by substituting information about a shape into the appropriate area formula. A formula may use letters instead of the words. For example, the area of a rectangle is found by multiplying the length times the width. This can be written A = *lw*. Below is a summary of the formulas you have learned to this point, using letters instead of words.

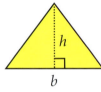

Triangle

$A = \frac{1}{2}bh$

Rectangle

$A = lw$

Square

$A = lw$ or $A = s^2$

EXAMPLE 1

Find the area of the triangle.

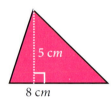

5 cm

8 cm

b = base
h = height

SOLUTION

Write the formula. $\text{Area} = \frac{1}{2}bh$

Substitute the given information. $\text{Area} = \frac{1}{2}(8)(5)$

Multiply. $\text{Area} = 20$

The area of the triangle is 20 square centimeters.

Composite figures are made up of two or more geometric shapes. Figures joined with shapes at their edges or with a part(s) removed are composite figures. The area of each shape in a composite figure must be determined in order to find the total area of the figure. Once the areas are calculated, decide whether to add or subtract the areas to find the overall area of the composite figure.

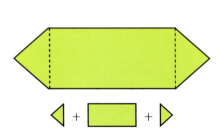

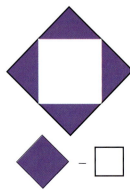

EXAMPLE 2

Calculate the area of the shaded region.

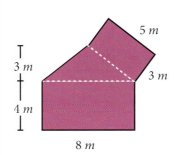

SOLUTION

Draw a diagram.

Find the area of each shape.

$$\text{Area} = lw = 8 \cdot 4 = 32 \ m^2$$

$$\text{Area} = \tfrac{1}{2}bh = \tfrac{1}{2}(8)(3) = 12 \ m^2$$

$$\text{Area} = lw = 5 \cdot 3 = 15 \ m^2$$

Add the areas of the three shapes.

$$\text{Area} = 32 + 12 + 15 = 59 \ m^2$$

The area of the shaded region is 59 square meters.

EXAMPLE 3

Calculate the area of the shaded region.

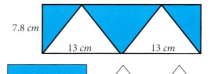

SOLUTION

Draw a diagram.

Find the area of the rectangle.

$$\text{Area} = lw = 26(7.8) = 202.8 \ cm^2$$

Find the area of the triangle.

$$\text{Area} = \tfrac{1}{2}bh = \tfrac{1}{2}(13)(7.8) = 50.7 \ cm^2$$

Subtract the area of the two triangles from the area of the rectangle.

$$\text{Area} = 202.8 - 50.7 - 50.7 = 101.4 \ cm^2$$

The area of the shaded region is 101.4 cm^2.

EXERCISES

Find the area of each figure.

1.

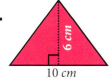

2.

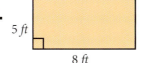

3.

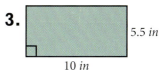

Find the area of each figure.

4.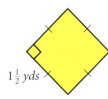

$1\frac{1}{2}$ yds

5.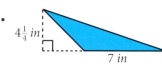

$4\frac{1}{4}$ in   7 in

6.

8.9 m

Sketch a diagram of each figure and label it with the given information. Calculate the area of each figure.

7. A triangle has a base of 6 *cm* and a height of 4 *cm*.

8. The length of a rectangle is 21 feet and the height is 9 feet.

9. The width of a rectangle is 4.5 inches. The length is 3.5 inches longer than the width.

10. The height of a triangle is $2\frac{1}{2}$ meters. Its base is $1\frac{1}{2}$ meters longer than the height.

Use a diagram to show how to find the area of each shaded region.

11.

12.

13.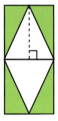

14.

Calculate the area of each shaded region.

15.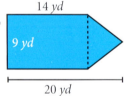

14 yd

9 yd

20 yd

16.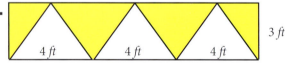

4 ft   4 ft   4 ft   3 ft

17.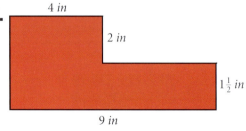

4 in

2 in

$1\frac{1}{2}$ in

9 in

18.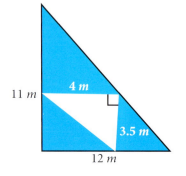

11 m   4 m

3.5 m

12 m

Calculate the area of each figure.

19.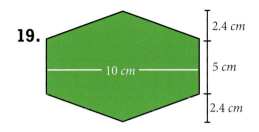

2.4 cm

10 cm

5 cm

2.4 cm

20.

8 in

4 in

10 in

21. The perimeter of a square is 72 *m*.
   a. What is the length of one side?
   b. Find the area of the square.

22. The length of a rectangle is 2.5 *cm*. The area is 20 *cm²*. What is the width of the rectangle?

23. Lois built a rectangular flower bed. She used four boards for the border of the flower bed. Two of the boards were 4 feet long. The other boards were 8 feet long.
   a. What was the area of the flower bed using these boards?
   b. She decided to cut each board in half to make two **square** flower beds. What was the area of each square flower bed?
   c. Did one or two flower beds give Lois more total area to plant flowers?

24. A window had 8 panes. Each pane measured 8 inches by 10 inches. What is the area of the entire window?

25. Sydney needs to cut rectangles from a 4 by 8 foot sheet of plywood. Each rectangle she cuts needs to be 1 foot by 3 feet. What is the maximum number of rectangles she can cut?

REVIEW

26. Sammi needs to cut a sheet of metal into 4 equal pieces to complete her art project. The sheet metal is 60.8 *cm* long. How long will each piece be?

27. Dave chose three lengths of cardboard to make a sign. The sign needs to measure 1 meter (100 *cm*) long. The three lengths of cardboard are 36.2 *cm*, 51.6 *cm* and 11.2 *cm*. Will he have enough cardboard to make the sign? Support your answer with words and/or mathematics.

Estimate each product or quotient using compatible numbers.

28. $19\frac{3}{4} \div 2\frac{1}{7}$

29. $25\frac{1}{4} \times 3\frac{1}{8}$

30. $73\frac{2}{3} \div 24\frac{1}{4}$

Draw nets for solids.
Find the surface area of a solid using its net.

The world is made up of many three-dimensional objects. From the blocks children play with to the Great Pyramid of Egypt, solids are all around you. A **solid** is a three-dimensional figure that encloses a part of space. In this lesson you will learn about five different types of solids. Many solids are made up of flat surfaces called **faces**. Each face is a circle or a **polygon** (a closed figure made up of three or more line segments). Solids may have one or two **bases**, often located on the top or bottom of a solid.

Name	Definition	Diagrams
Prism	A solid formed by two congruent, parallel bases and rectangular sides.	
Pyramid	A solid with a polygonal base and triangular sides that meet at a vertex.	
Cylinder	A solid formed by two congruent and parallel circular bases.	
Cone	A solid formed by one circular base and a curved surface which connects the base and the vertex.	
Sphere	A solid formed by a set of points in space that are the same distance from a center point.	

A **net** is a two-dimensional pattern that can be folded to form a three-dimensional figure. Below are some nets and their corresponding solids.

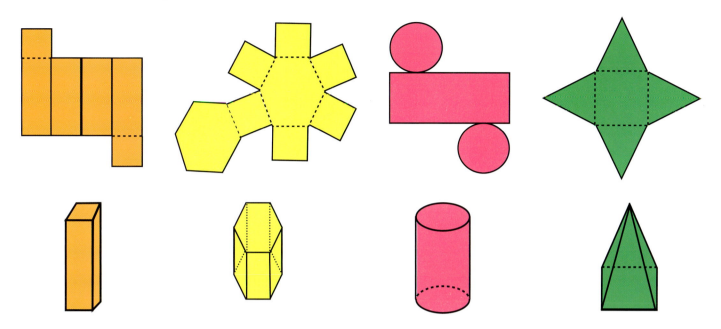

Prisms and pyramids are named using the name of the base before the word "prism" or "pyramid." For example, a prism whose base is a triangle is called a "triangular prism." A pyramid whose base is a rectangle is called a "rectangular pyramid."

EXAMPLE 1

Sketch a net for a triangular prism.

SOLUTION

Step 1: First sketch one base.

Step 2: Sketch one face that connects to the base.

Step 3: Sketch the other base.

Step 4: Finally, sketch the remaining faces.

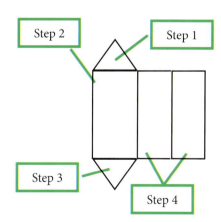

EXAMPLE 2

Sketch a net for a pentagonal pyramid.

SOLUTION

Start by sketching the base of the pyramid (a pentagon).

Sketch each face (triangle) by attaching it to one of the base edges.

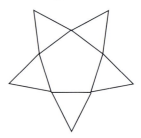

The **surface area** of a three-dimensional figure is the sum of the areas of all the surfaces. This includes the areas of the base(s) and sides. You can calculate the surface area of a figure by drawing its net, finding the area of each shape in the net and adding the areas together.

EXPLORE!   **NETTING A SOLID**

Locate a small rectangular prism, such as a juice box, to use in this activity.

Step 1: Trace the base of your prism near the edge of a large piece of paper.

Step 2: Tip your solid down on its side. Line up the solid so that the edge of the face touches the corresponding edge on the drawing. Trace this face.

Step 3: Roll the prism on the piece of paper and trace each side one at a time. Finally, stand the prism up and trace the remaining base. You should never lift the prism.

Step 4: Label each face as top, bottom or side. There will be more than one side. This is a net of your prism.

Step 5: Measure the height, width and length of your original prism to the nearest tenth of a centimeter. Record these measurements on your net.

Step 6: Find the area of each rectangle on your net. Add all the areas together to find the surface area of your figure.

Step 7: Greg measured a cereal box and found that it was 14 inches tall, 8 inches wide and 3 inches deep. Sketch a net of his cereal box. Your sketch does not need to be full size.

Step 8: Use your net to find the surface area of the cereal box.

FINDING THE SURFACE AREA OF A SOLID

1. Draw a net of the solid.
2. Record the measurements on the net.
3. Find the area of each part of the net.
4. Find the sum of all the areas and label your answer.

EXAMPLE 3

Find the surface area of the prism.

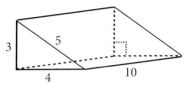

SOLUTION

Draw the net of the figure.

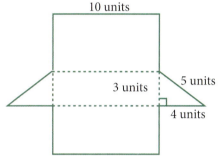

The two bases are congruent triangles.
Find the area of the triangle.

3 units | 5 units
4 units

$Area = \frac{1}{2} \times base \times height$

Substitute the base and the height.

$Area = \frac{1}{2}(4)(3)$

Multiply.

$Area = 6$ sq. units

Multiply by two for the area of both bases.

Area of both $= 2(6) = 12$ sq. units

Find the area of each rectangle.

10 units
5 units

$Area = length \times width$
$= 10(5)$
$= 50$ sq. units

10 units
3 units

$Area = length \times width$
$= 10(3)$
$= 30$ sq. units

10 units
4 units

$Area = length \times width$
$= 10(4)$
$= 40$ sq. units

Find the total area by adding the areas of the
two triangles and three rectangles.

$12 + 50 + 30 + 40 = 132$ sq. units

EXERCISES

Sketch a net of each solid.

1.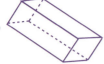

2.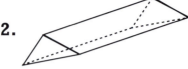

3.

4.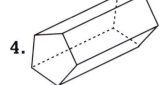

5.

6.

7.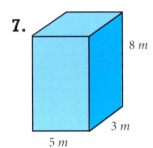

8 m

3 m

5 m

a. Draw a net of the prism at the left. Label the lengths of each side.
b. Find the area of each polygon in the net. Write the areas in the corresponding polygons.
c. Add all areas to find the total surface area.

8.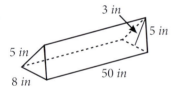

3 in

5 in

5 in

50 in

8 in

a. Draw a net of the prism at the left. Label the key information on each polygon.
b. Find the area of each polygon in the net. Write the areas in the corresponding polygons.
c. Add all areas to find the total surface area.

Find the surface area of each prism.

9.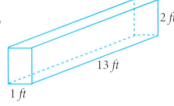

2 ft

13 ft

1 ft

10.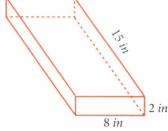

15 in

2 in

8 in

11.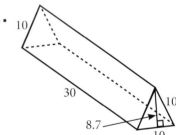

10

30

8.7

10

10

12. Sergio wrapped a gift for his mother. The box was a rectangular prism with dimensions of 12 inches by 10 inches by 4 inches.

    a. How much wrapping paper did Sergio need to exactly cover the box?

    b. Sergio ended up using 450 square inches of wrapping paper. Why do you think he used more than the answer in **part a**?

Find the surface area of each pyramid.

13.
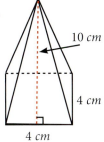
10 cm
4 cm
4 cm

   a. Draw a net of the pyramid.
   b. Find the area of each figure in the net.
   c. Find the surface area of the pyramid.

14.
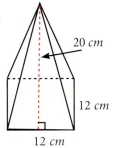
20 cm
12 cm
12 cm

   a. Draw a net of the pyramid.
   b. Find the area of each figure in the net.
   c. Find the surface area of the pyramid.

15.
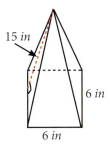
15 in
6 in
6 in

16.
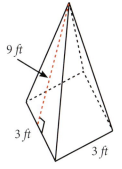
9 ft
3 ft
3 ft

17. Terry made game pieces in the shape of square pyramids. Each piece has a base edge of 2 *cm* and a slant height of 4 *cm*. He will paint all of the pieces. He needs to know how much paint he needs.
   a. Find the surface area of one game piece.
   b. Each game has 24 game pieces. Find the total surface area of one set of game pieces.
   c. He wants to make 12 games. What is the total surface area for all 12 games?
   d. A can of paint covers 400 square centimeters. How many cans of paint will he need?

REVIEW

Find the value of each expression. Write each answer in simplest form.

18. $2\frac{1}{2} \times 1\frac{3}{4}$

19. $5\frac{1}{4} \div 3\frac{1}{2}$

20. $1\frac{7}{8} + 1\frac{3}{4}$

21. $6\frac{1}{5} - 2\frac{1}{10}$

22. $7\frac{1}{6} \div 1\frac{2}{3}$

23. $4\frac{2}{3} \times 5\frac{1}{2}$

TIC-TAC-TOE ~ SIMILARITY SWAP

Draw three solids that have the same volume but different surface areas. Solids should all be labeled with their measurements.

Create three pairs of solids that have the same surface area but different volumes. Solids should all be labeled with their measurements.

Have another student or an adult find the surface area and volume of each solid shape to check your work.

TIC-TAC-TOE ~ VOCABULARY MEMORY

Copy each vocabulary word from the Blocks of this book onto separate cards. Copy the definition for each vocabulary word onto a different card of the same size. Create a memory game with the cards. The goal of the game is to match the word with the correct definition. Play this game in math class with your classmates to review vocabulary. Make a key showing the correct matches.

TIC-TAC-TOE ~ WHERE IN THE WORLD?

Find pictures of solids in the real world using magazines, real photos or pictures on the internet. Print or cut out the pictures. Include at least 8 solid figures of at least three types of solids.

Make a poster of these solids using the pictures you have cut out or printed showing where you can find solid figures in the real world.

Next to each photo or printed pictures, draw a net of that solid, if possible.

VOLUME WITH FRACTIONAL DIMENSIONS

 Find the volume of rectangular prisms with fractional side lengths.

The number of cubic units needed to fill a solid is called the **volume**. A rectangular prism is a solid that is formed by two congruent, parallel bases that are rectangles. All of the sides are rectangles also. All of the boxes shown below are rectangular prisms. Any side can be the length, width or height depending on the orientation of the box.

VOLUME OF A RECTANGULAR PRISM

The volume of a rectangular prism is found using the formula

$$V = lwh$$

where l = length, w = width and h = height.

When measuring objects in the real world, measurements are often not whole numbers. When finding the volume of a prism with fraction measurements, you must change all the whole numbers and mixed numbers to improper fractions before multiplying.

EXAMPLE 1 **Find the volume of the rectangular prism when $l = 2$ in, $w = 3\frac{1}{4}$ in and $h = 2\frac{1}{2}$ in.**

SOLUTION

Change all mixed and whole numbers to improper fractions.

$$l = \frac{2}{1}, \ w = \frac{13}{4}, \ h = \frac{5}{2}$$

Use the formula $V = lwh$.

$$V = \frac{2}{1} \times \frac{13}{4} \times \frac{5}{2}$$

Cross-simplify and multiply.

$$V = \frac{\overset{1}{\cancel{2}}}{1} \times \frac{13}{4} \times \frac{5}{\underset{1}{\cancel{2}}} = \frac{65}{4}$$

> Volume is labeled in cubic units.

Simplify.

$$V = \frac{65}{4} = 16\frac{1}{4} \ in^3$$

EXAMPLE 2

Storage Solutions, Inc. uses a box with a length of $3\frac{1}{4}$ feet, a width of $1\frac{1}{4}$ feet and a height of 8 feet. What is the volume of the box?

SOLUTION

Change all mixed and whole numbers to improper fractions.

$$l = \frac{13}{4}, \ w = \frac{5}{4}, \ h = \frac{8}{1}$$

Use the formula V = *lwh*.

$$V = \frac{13}{4} \times \frac{5}{4} \times \frac{8}{1}$$

Cross-simplify and multiply.

$$V = \frac{13}{\cancel{4}} \times \frac{5}{\cancel{4}} \times \frac{\cancel{8}^{1}}{1} = \frac{65}{2}$$

Simplify.

$$V = \frac{65}{2} = 32\frac{1}{2} \text{ cubic feet}$$

EXPLORE!   **MEASURING VOLUME**

Julie had three small boxes. She wanted to find the volume of each. She knew the width of each box. Measure the height and length of each box. Then find the volume of each box.

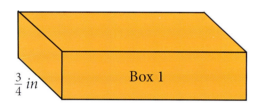 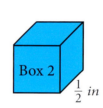 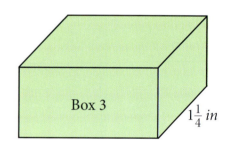

Step 1: Using a ruler, measure the length and height of each rectangular prism to the nearest quarter inch.

	Length	Width	Height
Box 1		$\frac{3}{4}$ in	
Box 2		$\frac{1}{2}$ in	
Box 3		$1\frac{1}{4}$ in	

Step 2: Find the volume of each box. Show all your work and label your answers.

Step 3: Which rectangular prism has the greatest volume?

Step 4: Julie wondered which box would increase the most in volume if she added 2 inches to the height. What would you tell her? Support your answer with words and/or symbols.

EXERCISES

Use the formula V = *lwh* to find the volume of a rectangular prism.

1. $l = \frac{1}{3}$ yd, $w = \frac{2}{5}$ yd, $h = \frac{15}{16}$ yd

2. $l = 1\frac{1}{3}$ in, $w = 3$ in, $h = 5\frac{1}{2}$ in

3. $l = 5\frac{1}{4}$ ft, $w = 2$ ft, $h = 2\frac{2}{3}$ ft

4. $l = 6$ units, $w = 2\frac{1}{4}$ units, $h = 4\frac{1}{2}$ units

5. $l = 1\frac{1}{5}$ in, $w = 3\frac{1}{3}$ in, $h = 5$ in

6. $l = 6\frac{2}{3}$ ft, $w = 1\frac{1}{2}$ ft, $h = 3\frac{3}{4}$ ft

7. A shed is $6\frac{2}{3}$ yards long, $9\frac{3}{4}$ yards tall and $4\frac{1}{3}$ yards deep.
    a. What is the volume of the shed?
    b. Jake had 300 cubic yards of hay bales to put in the shed. Will the hay bales fit?

Measure the following rectangular prisms to the nearest quarter inch. Find the volume of each prism. The width is given.

8.

$\frac{1}{2}$ in

9.

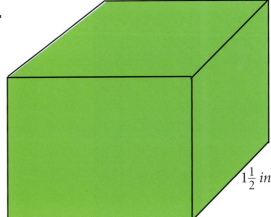

$1\frac{1}{2}$ in

10.

1 in

11. Laura filled a box with packing peanuts. The box was $3\frac{1}{2}$ feet long, $2\frac{1}{2}$ feet wide and 2 feet tall. How many cubic feet did Laura fill?

12. A cereal box was $1\frac{1}{2}$ feet tall, $1\frac{1}{12}$ feet long and $\frac{1}{6}$ foot wide. How many cubic feet of cereal can the box hold?

13. Mike filled a planter box with potting soil. The box was a cube that was $\frac{1}{3}$ yard wide. How many cubic yards of potting soil did Mike put in the planter box?

14. Carlie poured rice into a storage box. The box measured $\frac{3}{4}$ foot tall, $\frac{1}{2}$ foot long and $\frac{1}{4}$ foot wide. How many cubic feet of rice filled the storage box?

15. Jenna has a cube that measures $\frac{2}{3}$ inch on each side.

    a. What is the volume of the cube?

    b. Ivan has a cube that has side lengths that are twice as long as Jenna's cube. Will his volume be twice as much? Support your answer with mathematics.

The volume of any prism can be found by multiplying the area of the base (*B*) by the height (*h*) of the prism using the formula V = *Bh*. The bases on a prism are parallel and congruent. Find the volume of each solid.

16. The area of the base on a hexagonal prism is 12 square inches. The height is $8\frac{5}{6}$ inches. What is the volume of the prism?

17. A pentagonal prism is $1\frac{1}{10}$ meters tall and has a base area of $6\frac{7}{10}$ square meters. What is the volume of the prism?

18. Base area = 22.5 *in²*

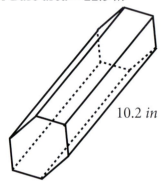

10.2 *in*

19.

5.4 *in*

6.5 *in*

11.75 *in*

20.

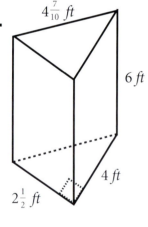

$4\frac{7}{10}$ *ft*

6 *ft*

4 *ft*

$2\frac{1}{2}$ *ft*

REVIEW

Find the value of each expression. Write each answer in simplest form.

21. $\frac{1}{3} \times \frac{5}{6}$

22. $\frac{7}{8} - \frac{1}{2}$

23. $\frac{10}{13} \times \frac{2}{5}$

24. $\frac{14}{15} \div \frac{2}{3}$

25. $\frac{9}{10} \div \frac{3}{5}$

26. $\frac{1}{4} + \frac{2}{3}$

Find the area of each polygon.

27.  $2\frac{1}{4}$ in

28. 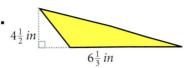 $4\frac{1}{2}$ in $6\frac{1}{3}$ in

29. 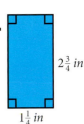 $2\frac{3}{4}$ in $1\frac{1}{4}$ in

---

TIC-TAC-TOE ~ NET PATTERNS

Draw net patterns for the following solids:

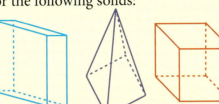

Sketch net patterns using solid lines to show where you cut. Use dotted lines to show where you fold.

Test your nets by tracing and cutting them out using your pattern as a guide. Build the solid using your net pattern and taping where you connect any two edges.

Write directions so that anyone could make the solid by using the pattern you created.

TIC-TAC-TOE ~ COMPOSITE FIGURES

Design four composite figures made up of rectangls, triangles and/or squares that are completely shaded. Give length measurements. Find the area of each composite figure.

Example:

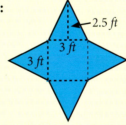 2.5 ft 3 ft 3 ft

Design four composite figures that have some shaded and some unshaded regions. Give length measurements. Find the area of the shaded region.

Example:

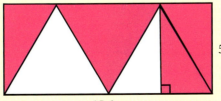 5.2 cm 15.6 cm

Have another student or an adult find the area of each composite figure to check your work.

Vocabulary

area	faces	solid
bases	net	surface area
composite figures	perpendicular	volume
	polygon	

 Calculate areas of rectangles, squares and triangles with lengths of sides that are fractions or mixed numbers.
Calculate the perimeter and area of squares, rectangles and triangles using the metric system.
Find the areas of composite figures by breaking them down into known shapes.
Draw nets for solids.
Find the surface area of a solid using its net.
Find the volume of rectangular prisms with fractional sides lengths.

Lesson 6.1 ~ Area with Fractions

Use the given measurements to find the area of the following shapes.

1. $2\frac{1}{4}$ in
$4\frac{1}{2}$ in

2. SQUARE $3\frac{1}{2}$ units

3. 3 in $4\frac{1}{2}$ in

4. $3\frac{1}{3}$ units $4\frac{1}{2}$ units

5. $6\frac{1}{5}$ units $4\frac{3}{8}$ units

6. SQUARE $4\frac{3}{4}$ in

Measure the sides of the following polygons, using a customary ruler, to the nearest quarter of an inch. Find the area of each polygon.

7.

8.

9.

Use the given measurements to find each perimeter and area.

10.  3.2 cm

4.1 cm

11.  7.5 m

12.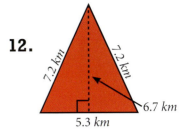

7.2 km   7.2 km

6.7 km

5.3 km

Measure the necessary lengths to the nearest tenth of a centimeter. Find the perimeter and area of each polygon.

13.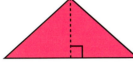

14.

15.

Lesson 6.3 ~ Areas of Composite Figures

● ●

Use a diagram or words to explain how to find the area of each shaded region.

16.

17.

Find the area of each figure.

18.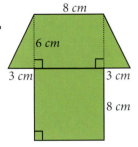

8 cm

6 cm

3 cm   3 cm

8 cm

19.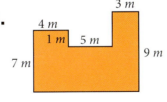

3 m

4 m

1 m   5 m

7 m   9 m

Calculate the area of each shaded region.

20.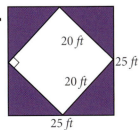

20 ft

25 ft

20 ft

25 ft

21.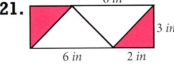

6 in

3 in

6 in   2 in

Sketch a net of each solid.

22.

23.

Find the surface area of each figure.

24.

2 ft

4 ft

15 ft

25.

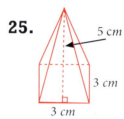

5 cm

3 cm

3 cm

26. Nina has a box of greeting cards that is 3 inches tall, 6.5 inches wide and 9 inches long. What is the surface area of the box?

Lesson 6.5 ~ Volume with Fractional Dimensions

Use the formula $V = lwh$ **to find the volume of each solid.**

27.

$\frac{3}{4}$ in

$1\frac{1}{2}$ in

$\frac{3}{4}$ in

28.

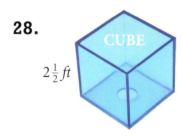

CUBE

$2\frac{1}{2}$ ft

29. $l = 3$ yds, $w = 2\frac{1}{3}$ yds, $h = 5\frac{1}{2}$ yds

30. $l = 1\frac{1}{2}$ ft, $w = 2$ ft, $h = 4\frac{1}{4}$ ft

31. Jaliene filled a frame with cement. The frame was $2\frac{2}{3}$ yards long, $3\frac{1}{2}$ yards wide and $\frac{1}{3}$ yard tall. How many cubic yards of cement did she use to fill the frame?

32. Tory has a cube that measures $6\frac{1}{2}$ inches on each side. What is the volume of his cube?

33. Tim poured wheat into a storage bin. The bin measured $3\frac{1}{2}$ feet long, $2\frac{1}{8}$ feet wide and 4 feet tall. How many cubic feet of wheat did he pour into the bin to completely fill it?

TIC-TAC-TOE ~ BOX BLUES

A cereal company is coming out with six new cereals. They have asked you to calculate costs for each of the six new boxes that will be made with cardboard. You are also in charge of figuring out the maximum amount of cereal that can be put into each box.

> Cardboard to make Boxes 1 and 2 costs $0.001 per square inch.
> Cereal to fill Boxes 1 and 2 costs $0.002 per cubic inch.

Box 1: $l = 9\frac{1}{2}$ in, $w = 3$ in, $h = 15\frac{1}{2}$ in   **Box 2:** $l = 9$ in, $w = 3.25$ in, $h = 14.5$ in

> Cardboard to make Boxes 3 and 4 costs $0.001 per square centimeter.
> Cereal to fill Boxes 3 and 4 costs $0.002 per cubic centimeter.

Box 3: $l = 22$ cm, $w = 5.6$ cm, $h = 45.3$ cm   **Box 4:** $l = 25.3$ cm, $w = 6.5$ cm, $h = 42$ cm

> Cardboard to make Boxes 5 and 6 costs $10.10 per square foot.
> Cereal to fill Boxes 5 and 6 costs $2.00 per cubic centimeter.

Box 5: $l = 1\frac{1}{4}$ ft, $w = \frac{1}{4}$ ft, $h = 1\frac{1}{2}$ ft   **Box 6:** $l = 1\frac{1}{3}$ ft, $w = \frac{1}{3}$ ft, $h = 2$ ft

Step 1: Sketch the net for each box. Label each side with the appropriate measurement.

Step 2: Calculate the surface area of each box. This will tell you how much cardboard the company will need.

Step 3: Calculate the cost to make each box using prices given.

Step 4: Calculate the volume of each box. This will tell you the maximum cubic units of cereal that can fill each box. Calculate the cost to fill each box using prices given.

Step 5: Which box of cereal will cost the company the least to make and fill?

Step 6: Which box of cereal will cost the company the most to make and fill?

CAREER FOCUS

ANA
MIDDLE SCHOOL PRINCIPAL

I am a middle school principal. My job has lots of different duties. One thing I do is make sure that my school is a safe and positive place for students and staff. I work with parents, students and teachers to make sure our students are achieving as highly as they can. My job includes managing our school budget and making sure we are doing everything the state and federal governments require.

Math is an important part of running any school. We must account for and keep track of money grants we receive. Teachers and other staff are paid salaries for which I must budget. Textbooks and supplies must be purchased. My school has to manage the budget in a way that best helps students learn. Student achievement is tracked using percentages. I regularly use math to see if students are making good progress toward our learning goals.

I completed a college program and got a master's degree to become a middle school principal. I also had to pass tests to make sure that I knew everything that would be required of a principal. Principals are licensed by the state. I have to renew my license every few years and keep current on what is happening in education. I am always learning new things about education as a principal.

A principal's salary can range from $65,000 - $90,000 per year. Principals also get other benefits like health insurance. Salaries depend on where in the state you work, what level of school you are the principal of and how many years experience you have.

I enjoy my profession because it allows me to have an impact on student learning and student success.

ACKNOWLEDGEMENTS

All Photos and Clipart ©2008 Jupiterimages Corporation and Clipart.com
with the exception of the cover photo and the following photos:

Decimals & Fractions Page 46
©iStockphoto.com/Ruth Black

Decimals & Fractions Page 57
©iStockphoto.com/Andres Balacazar

Decimals & Fractions Page 50
©iStockphoto.com/matka_wariatka

Decimals & Fractions Page 52
©iStockphoto.com/jurate

Decimals & Fractions Page 82
©iStockphoto.com/Lisa F. Young

Decimals & Fractions Page 123
©iStockphoto.com/ericsphotography

Decimals & Fractions Page 184
©iStockphoto.com/nilsz

Decimals & Fractions Page 189
©iStockphoto.com/malerapaso

Decimals & Fractions Page 192
©iStockphoto.com/mocker_bat

Layout and Design by Judy St. Lawrence

Cover Design by Schuyler St. Lawrence

Glossary Translation by Heather Contreras

CORE FOCUS ON MATH
GLOSSARY ~ GLOSARIO

A

Absolute Value	The distance a number is from 0 on a number line.	**Valor Absoluto**	La distancia de un número desde el 0 en una recta numérica.
Acute Angle	An angle that measures more than 0° but less than 90°.	**Ángulo Agudo**	Un ángulo que mide mas 0° pero menos de 90°.
Adjacent Angles	Two angles that share a ray.	**Ángulos Adyacentes**	Dos ángulos que comparten un rayo.
Algebraic Expression	An expression that contains numbers, operations and variables.	**Expresiones Algebraicas**	Una expresión que contiene números, operaciones y variables.
Alternate Exterior Angles	Two angles that are on the outside of two lines and are on opposite sides of a transversal.	**Ángulos Exteriores Alternos**	Dos ángulos que están afuera de dos rectas y están a lados opuestos de una transversal.
Alternate Interior Angles	Two angles that are on the inside of two lines and are on opposites sides of a transversal.	**Ángulos Interiores Alternos**	Dos ángulos que están adentro de dos rectas y están a lados opuestos de una transversal.
Angle	A figure formed by two rays with a common endpoint.	**Ángulo**	Una figura formada por dos rayos con un punto final en común.

| Area | The number of square units needed to cover a surface. | Área | El número de unidades cuadradas necesitadas para cubrir una superficie. |

| Ascending Order | Numbers arranged from least to greatest. | Progresión Ascendente | Los números ordenados de menor a mayor. |

| Associative Property | A property that states that numbers in addition or multiplication expressions can be grouped without affecting the value of the expression. | Propiedad Asociativa | Una propiedad que establece que los números en expresiones de suma o de multiplicación pueden ser agrupados sin afectar el valor de la expresión. |

| Axes | A horizontal and vertical number line on a coordinate plane. | Ejes | Una recta numérica horizontal y vertical en un plano de coordenadas. |

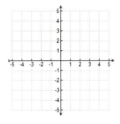 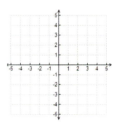

| Axis of Symmetry | The line of symmetry on a parabola that goes through the vertex. | El Eje De Las Simetría | La linia de simetría de una parábola que pasa por el vértice. |

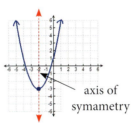
axis of symametry

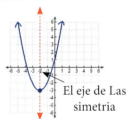
El eje de Las simetria

B

| Bar Graph | A graph that uses bars to compare the quantities in a categorical data set. | Gráfico de Barras | Una gráfica que utiliza barras para comparar las cantidades en un conjunto de datos categórico. |

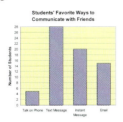
Students' Favorite Ways to Communicate with Friends

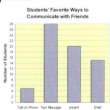
Students' Favorite Ways to Communicate with Friends

| Base (of a power) | The base of the power is the repeated factor. In x^a, x is the base. | Base (de la potencia) | La base de la potenciación es el factor repatidio. En x^a, x es la base. |

Base (of a solid)	See Prism, Cylinder, Pyramid and Cone.	Base (de un sólido)	Ver Prisma, Cilindro, Pirámide y Cono.
Base (of a triangle)	Any side of a triangle.	Base (de un triángulo)	Cualquier lado de un triángulo.
Bias	A problem when gathering data that affects the results of the data.	Sesgo	Un problema que ocurre cuando se recogen datos que afectan los resultados de los datos.
Biased Sample	A group from a population that does not accurately represent the entire population.	Muestra Sesgada	Un grupo de una población que no representa con exactitud la población entera.
Binomials	Expressions involving two terms (i.e. $x - 2$).	Binomiales	Expresiones que impliquen dos terminos. (es decir: $x - 2$).
Bivariate Data	Data that describes two variables and looks at the relationship between the two variables.	Datos de dos Variables	Los datos que describen dos variables y analiza la relación entre estas dos variables.
Box-and-Whisker Plot	A diagram used to display the five-number summary of a data set.	Diagrama de Líneas y Bloques	Un diagrama utilizado para mostrar el resumen de cinco números de un conjunto de datos.

C

Categorical Data	Data collected in the form of words.	Datos Categóricos	Datos recopilados en la forma de palabras.
Center of a Circle	The point inside a circle that is the same distance from all points on the circle.	Centro de un Círculo	Un ángulo dentro de un círculo que está a la misma distancia de todos los puntos en el círculo.

Central Angle	An angle in a circle with its vertex at the center of a circle.	Ángulo Central	Un ángulo en un círculo con su vértice en el centro del círculo.
Chord	A line segment with endpoints on a circle.	Cuerda	Un segmento de la recta con puntos finales en el círculo.
Circle	The set of all points that are the same distance from a center point.	Círculo	El conjunto de todos los puntos que están a la misma distancia de un punto central.
Circumference	The distance around a circle.	Circunferencia	La distancia alrededor de un círculo.
Coefficient	The number multiplied by a variable in a term.	Coeficiente	El número multiplicado por una variable en un término.
Commutative Property	A property that states numbers can be added or multiplied in any order.	Propiedad Conmutativa	Una propiedad que establece que los números pueden ser sumados o multiplicados en cualquier orden.
Compatible Numbers	Numbers that are easy to mentally compute; used when estimating products and quotients.	Números Compatibles	Números que son fáciles de calcular mentalmente; utilizado cuando se estiman productos y cocientes.
Complementary Angles	Two angles whose sum is 90°.	Ángulos Complementarios	Dos ángulos cuya suma es de 90°.
Complements	Two probabilities whose sum is 1. Together they make up all the possible outcomes without repeating any outcomes.	Complementos	Dos probabilidades cuya suma es de 1. Juntos crean todos los posibles resultados sin repetir alguno.

Completing the Square	The creation of a perfect square trinomial by adding a constant to an expression in the form $x^2 + bx$.	Terminado el Cuadrado	La creación de un trinomio cuadrado perfecto por adición de una constante a una expresión en la forma $x^2 + bx$.
Complex Fraction	A fraction that contains a fractional expression in its numerator, denominator or both. $$\dfrac{\frac{3}{4}}{\frac{3}{8}}$$	Fracción Compleja	Una fracción que contiene una expresión fraccionaria en su numerador, el denominador o ambos. $$\dfrac{\frac{3}{4}}{\frac{3}{8}}$$
Composite Figure	A geometric figure made of two or more geometric shapes.	Figura Compuesta	Una figura geométrica formada por dos o más formas geométricas.
Composite Number	A whole number larger than 1 that has more than two factors.	Número Compuesto	Un número entero mayor que el 1 con más de dos factores.
Composite Solid	A solid made of two or more three-dimensional geometric figures.	Sólido Compuesto	Un sólido formado por dos o más figuras geométricas tridimensionales.
Composition of Transformations	A series of transformations on a point.	Composición de Transformaciones	Una serie de transformaciones en un punto.
Compound Probability	The probability of two or more events occurring.	Compuesto de Probabilidad	La probabilidad de dos o más eventos ocurriendo.
Conditional Frequency	The ratio of the observed frequency to the total number of frequencies in a given category from an experiment or survey.	Frecuencia Condicional	La relación de una frecuencia observada para el número total de frecuencias en una categoría dada del experimento o encuesta.
Cone	A solid formed by one circular base and a vertex.	Cono	Un sólido formado por una base circular y una vértice.
Congruent	Equal in measure.	Congruente	Igual en medida.

| **Congruent Figures** | Two shapes that have the exact same shape and the exact same size. | **Figuras Congruentes** | Dos figuras que tienen exactamente la misma forma y el mismo tamaño. |

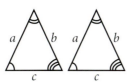 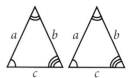

| **Constant** | A term that has no variable. | **Constante** | Un término que no tiene variable. |

| **Continuous** | When a graph can be drawn from beginning to end without any breaks. | **Continuo** | Cuando una gráfica puede ser dibujada desde principio a fin sin ninguna interrupción. |

| **Conversion** | The process of renaming a measurement using different units. | **Conversión** | El proceso de renombrar una medida utilizando diferentes unidades. |

| **Coordinate Plane** | A plane created by two number lines intersecting at a 90° angle. | **Plano de Coordenadas** | Un plano creado por dos rectas numéricas que se intersecan a un ángulo de 90°. |

 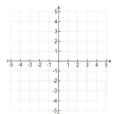

| **Correlation** | The relationship between two variables in a scatter plot. | **Correlación** | La relación entre dos variables en un gráfico de dispersión. |

| **Corresponding Angles** | Two non-adjacent angles that are on the same side of a transversal with one angle inside the two lines and the other on the outside of the two lines. | **Ángulos Correspondientes** | Dos ángulos no adyacentes que están en el mismo lado de una transversal con un ángulo adentro de las dos rectas y el otro afuera de las dos rectas. |

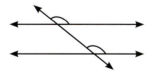 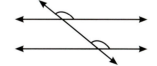

| **Corresponding Parts** | The angles and sides in similar or congruent figures that match. | **Partes Correspondientes** | Los ángulos y lados en figuras similares o congruentes que concuerdan. |

Cube Root	One of the three equal factors of a number.	Raíz Cúbica	Uno de los tres factores iguales de un número.
	$3 \cdot 3 \cdot 3 = 27 \quad \sqrt[3]{27} = 3$		$3 \cdot 3 \cdot 3 = 27 \quad \sqrt[3]{27} = 3$
Cubed	A term raised to the power of 3.	Cubicado	Un término elevado a la potencia de 3.
Cylinder	A solid formed by two congruent and parallel circular bases.	Cilindro	Un sólido formado por dos bases circulares congruentes y paralelas.

D

Decimal	A number with a digit in the tenths place, hundredths place, etc.	Decimal	Un número con un dígito en las décimas, las centenas, etc.
Degrees	A unit used to measure angles.	Grados	Una unidad utilizada para medir ángulos.
Dependent Events	Two (or more) events such that the outcome of one event affects the outcome of the other event(s).	Eventos Dependiente	Dos (o más) eventos de tal manera que el resultado de un evento afecta el resultado del otro evento (s).
Dependent Variable	The variable in a relationship that depends on the value of the independent variable.	Variable Dependiente	La variable en una relación que depende del valor de la variable independiente.

Descending Order	Numbers arranged from greatest to least.	Progresión Descendente	Los números ordenados de mayor a menor.

Diameter	The distance across a circle through the center.	**Diámetro**	La distancia a través de un círculo por el centro.
Dilation	A transformation which changes the size of the figure but not the shape. 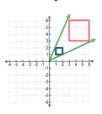	**Dilatación**	Una transformación que cambia el tamaño de la figura, pero no la forma. 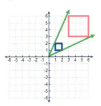
Direct Variation	A linear function that passes through the origin and has the equation $y = mx$.	**Variación Directa**	Una función lineal que pasa a través del origen y tiene la ecuación $y = mx$.
Discount	The decrease in the price of an item.	**Descuento**	La disminución de precio en un artículo.
Discrete	When a graph can be represented by a unique set of points rather than a continuous line.	**Discreto**	Cuando una gráfica puede ser representada por un conjunto de puntos único en vez de una recta continua.
Discriminant	In the quadratic formula, the expression under the radical sign. The discriminant provides information about the number of real roots or solutions of a quadratic equation. $$\frac{-b \pm \sqrt{b^2 - 4ac}}{2a}$$	**Discriminante**	En la fórmula cuadrática, la expresión bajo el signo radical. El discriminante proporciona información sobre el número o las verdaderas raíces o soluciones de una ecuación cuadrática. $$\frac{-b \pm \sqrt{b^2 - 4ac}}{2a}$$
Distance Formula	A formula used to find the distance between two points on the coordinate plane. $$d = \sqrt{(x_2 - x_1)^2 + (y_2 - y_1)^2}$$	**Fórmula de Distancia**	Una fórmula utilizada para encontrar la distancia entre dos puntos en un plano de coordenadas. $$d = \sqrt{(x_2 - x_1)^2 + (y_2 - y_1)^2}$$

Distributive Property	A property that can be used to rewrite an expression without parentheses. $$a(b + c) = ab + ac$$	Propiedad Distributiva	Una propiedad que puede ser utilizada para reescribir una expresión sin paréntesis: $$a(b + c) = ab + ac$$
Dividend	The number being divided. $$\textbf{100} \div 4 = 25$$	Dividendo	El número que es dividido. $$\textbf{100} \div 4 = 25$$
Divisor	The number used to divide. $$100 \div \textbf{4} = 25$$	Divisor	El número utilizado para dividir. $$100 \div \textbf{4} = 25$$
Domain	The set of input values of a function.	Dominio	El conjunto de valores entrados de la función.
Dot Plot	A data display that consists of a number line with dots equally spaced above data values. 	Punto de Gráfico	Una visualización de datos que consiste de una línea numérica con puntos igualmente espaciados sobre valores de datos. 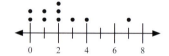
Double Stem-and-Leaf Plot	A stem-and-leaf plot where one set of data is placed on the right side of the stem and another is placed on the left of the stem. 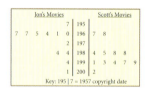	Doble Gráfica de Tallo y Hoja	Una gráfica de tallo y hoja donde un conjunto de datos es colocado al lado derecho del tallo y el otro es colocado a la izquierda del tallo.

E

Edge	The segment where two faces of a solid meet.	Arista (Borde)	El segmento donde dos caras de un sólido se encuentran.

English		Español	
Elimination Method	A method for solving a system of linear equations.	**Método de Eliminación**	Un método para resolver un sistema de ecuaciones lineales.
Enlargement	A dilation that creates an image larger than its pre-image.	**Agrandamiento**	Una dilatación que crea una imagen más grande que su pre-imagen.
Equally Likely	Two or more possible outcomes of a given situation that have the same probability.	**Igualmente Probables**	Dos o más posibles resultados de una situación dada que tienen la misma probabilidad.
Equation	A mathematical sentence that contains an equals sign between 2 expressions.	**Ecuación**	Una oración matemática que contiene un símbolo de igualdad entre dos expresiones.
Equiangular	A polygon in which all angles are congruent.	**Equiángulo**	Un polígono en el cual todos los ángulos son congruentes.
Equilateral	A polygon in which all sides are congruent.	**Equilátero**	Un polígono en el cual todos los lados son congruentes.
Equivalent Decimals	Two or more decimals that represent the same number.	**Decimales Equivalentes**	Dos o más decimales que representan el mismo número.
Equivalent Expressions	Two or more expressions that represent the same algebraic expression.	**Expresiones Equivalentes**	Dos o más expresiones que representan la misma expresión algebraica.
Equivalent Fractions	Two or more fractions that represent the same number.	**Fracciones Equivalentes**	Dos o más fracciones que representan el mismo número.
Evaluate	To find the value of an expression.	**Evaluar**	Encontrar el valor de una expresión.
Even Distribution	A set of data values that is evenly spread across the range of the data.	**Distribución Igualada**	Un conjunto de valores de datos que es esparcido de modo uniforme a través del rango de los datos.

English		Spanish	
Event	A desired outcome or group of outcomes.	Evento	Un resultado o grupo de resultados deseados.
Experimental Probability	The ratio of the number of times an event occurs to the total number of trials.	Probabilidad Experimental	La razón de la cantidad de veces que un suceso ocurre a la cantidad total de intentos.
Exponent	In x^a, a is the exponent. The exponent shows the number of times the factor (x) is repeated.	Exponente	En x^a, a es el exponente. El exponente indica el número de veces que se repite el factor (x).
Exponential Function	A function that can be described by an equation in the form $f(x) = bm^x$.	Función Exponencial	Una función que puede ser descrito por una ecuasión en la forma $f(x) = bm^x$.

F

Face	A polygon that is a side or base of a solid.	Cara	Un polígono que es una base de lado de un sólido.

face

cara

Factors	Whole numbers that can be multiplied together to find a product.	Factores	Números enteros que pueden ser multiplicados entre si para encontrar un producto.
First Quartile (Q1)	The median of the lower half of a data set.	Primer Cuartil (Q1)	Mediana de la parte inferior de un conjunto de datos.
Five-Number Summary	Describes the spread of a data set using the minimum, 1st quartile, median, 3rd quartile and maximum.	Sumario de Cinco Números	Describe la extensión de un conjunto de datos utilizando el mínimo, el primer cuartil, la mediana el tercer cuartil y el máximo.
Formula	An algebraic equation that shows the relationship among specific quantities.	Fórmula	Una ecuación algebraica que enseña la relación entre cantidades específicas.
Fraction	A number that represents a part of a whole number, written as $\frac{numerator}{denominator}$.	Fracción	Un número que representa una parte de un número entero, escrito como $\frac{numerador}{denominador}$.

| Frequency | The number of times an item occurs in a data set. | Frecuencia | La cantidad de veces que un artículo ocurre en un conjunto de datos. |

| Frequency Table | A table which shows how many times a value occurs in a given interval. | Tabla de Frecuencia | Una tabla que enseña cuantas veces un valor ocurre en un intervalo dado. |

Weight of Newborn (in Pounds)	Tally
4 – 5.5		
5.5 – 7	|||
7 – 8.5	|||||
8.5 – 10	||
10 – 11.5		

Weight of Newborn (in Pounds)	Tally
4 – 5.5		
5.5 – 7	|||
7 – 8.5	|||||
8.5 – 10	||
10 – 11.5		

| Function | A relationship between two variables that has one output value for each input value. | Función | Una relación entre dos variables que tiene un valor de salida para cada valor de entrada. |

G

| General Form | A quadratic function is in general form when written $f(x) = ax^2 + bx + c$ where $a \neq 0$. | Forma General | Una función cuadrática es en forma general cuándo escrito $f(x) = ax^2 + bx + c$ donde $a \neq 0$. |

| Geometric Probability | Ratios of lengths or areas used to find the likelihood of an event. | Probabilidad Geométrica | Razones de longitudes o áreas utilizadas para encontrar la probabilidad de un suceso. |

| Geometric Sequence | A list of numbers that begins with a starting value. Each term in the sequence is generated by multiplying the previous term in the sequence by a constant multiplier. | Secuenciación Geométrica | Una lista de números que comienza con un valor inicial. Cada término de la secuencia se genera al multiplicar el término anterior de la secuencia por un multiplicar constante. |

| Greatest Common Factor (GCF) | The greatest factor that is common to two or more numbers. | Máximo Común Divisor (MCD) | El máximo divisor que le es común a dos o más números. |

| Grouping Symbols | Symbols such as parentheses or fraction bars that group parts of an expression. | Símbolos de Agrupación | Símbolos como el paréntesis o barras de fracción que agrupan las partes de una expresión. |

H

Height of a Triangle | A perpendicular line drawn from the side whose length is the base to the opposite vertex.

Altura de un Triángulo | Una recta perpendicular dibujada desde el lado cuya longitud es la base al vértice opuesto.

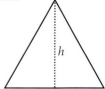

Histogram | A bar graph that displays the frequency of numerical data in equal-sized intervals.

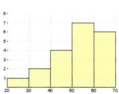

Histograma | Un gráfico de barras que muestra la frecuencia de datos numéricos en intervalos de tamaños iguales.

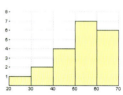

Hypotenuse | The side opposite the right angle in a right triangle.

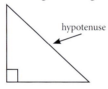

Hipotenusa | El lado opuesto el ángulo recto en un triángulo rectángulo.

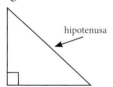

I-J-K

Image | A point or figure which is the result of a transformation or series of transformations.

Imagen | Un punto o figura que es el resultado de una transformación o una serie de transformaciones.

Improper Fraction | A fraction whose numerator is greater than or equal to its denominator.

Fracción Impropia | Una fracción cuyo numerador es mayor o igual a su denominador.

Independent Events | Two (or more) events such that the outcome of one event does not affect the outcome of the other event(s).

Eventos Independientes | Dos (o más) eventos de tal manera que el resultado de un evento no afecta el resultado del otro evento (s).

Independent Variable	The variable representing the input values.	Variable Independiente	La variable que representa los valores entratos.

independent

independiente

Inequality	A mathematical sentence using $<$, $>$, $\leq$ or $\geq$ to compare two quantities.	Desigualdad	Un enunciado matemático usando $<$, $>$, $\leq$ ó $\geq$ para comparar dos cantidades.
Inference	A logical conclusion based on known information.	Inferencia	Una conclusión lógica basada en la información conocida.
Input-Output Table	A table used to describe a function by listing input values with their output values.	Tabla de Entrada y Salida	Una tabla utilizada para describir una función al enumerar valores de entrada con sus valores de salidas.

Input, x	Output, y

Input, x	Output, y

Integers	The set of all whole numbers, their opposites, and 0.	Enteros	El conjunto de todos los números enteros, sus opuestos y 0.
Interquartile Range (IQR)	The difference between the 3rd quartile and the 1st quartile in a set of data.	Rango Intercuartil (IQR)	La diferencia entre el tercer cuartil y el primer cuartil en un conjunto de datos.
Inverse Operations	Operations that undo each other.	Operaciones Inversas	Operaciones que se cancelan la una a la otra.
IQR Method	A method for determining outliers using interquartile ranges.	Método IQR	Un método para determinar los datos aberrantes.
Irrational Numbers	A number that cannot be expressed as a fraction of two integers.	Números Irracionales	Un número que no puede ser expresado como una fracción de dos enteros.

Isosceles Trapezoid	A trapezoid that has congruent legs.	Trapezoide Isósceles	Un trapezoide con catetos congruentes.

Isosceles Triangle	A triangle that has two or more congruent sides.	Triángulo Isósceles	Un triángulo que tiene dos o más lados congruentes.

L

Lateral Face	A side of a solid that is not a base.	Cara Lateral	Un lado de un sólido que no sea una base.
Least Common Denominator (LCD)	The least common multiple of two or more denominators.	Mínimo Común Denominador (MCD)	El mínimo común múltiplo de dos o más denominadores.
Least Common Multiple (LCM)	The smallest nonzero multiple that is common to two or more numbers.	Mínimo Común Múltiplo (MCM)	El múltiplo más pequeño que no sea cero que le es común a dos o más números.
Leg	The two sides of a right triangle that form a right angle.	Cateto	Los dos lados de un triángulo rectángulo que forman un ángulo recto.
	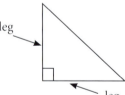		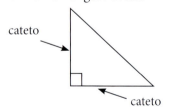
Like Terms	Terms that have the same variable(s).	Términos Semejantes	Términos que tienen el mismo variable(s).

English		Spanish	
Line of Best Fit	A line which best represents the pattern of a two-variable data set.	**Recta de Mejor Ajuste**	Una recta que mejor representa el patrón de un conjunto de datos de dos variables.
Linear Equation	An equation whose graph is a line.	**Ecuación Lineal**	Una ecuación cuya gráfica es una recta.
Linear Function	A function whose graph is a line.	**Función Lineal**	Una función cuya gráfica es una recta.
Linear Pair	Two adjacent angles whose non-common sides are opposite rays.	**Par Lineal**	Dos ángulos adyacentes cuyos lados no comunes son rayos opuestos.

M

Markup	The increase in the price of an item.	**Margen de Beneficio**	El aumento de precio en un artículo.
Maximum	The highest point on a curve.	**Máximo**	El punto más alto en la curva.
Mean	The sum of all values in a data set divided by the number of values.	**Media**	La suma de todos los valores en un conjunto de datos dividido entre la cantidad de valores.
Mean Absolute Deviation	A statistic that shows the average distance from the mean for all numbers in a data set.	**Desviación Media Absoluta**	Una estadística que muestra la distancia promedio entre la media de todos los números en una serie de datos.

English		Español	
Measures of Center	Numbers that are used to represent a data set with a single value; the mean, median, and mode are the measures of center.	**Medidas de Centro**	Números que son utilizados para representar un conjunto de datos con un solo valor; la media, la mediana, y la moda son las medidas de centro.
Measures of Variability	Statistics that help determine the spread of numbers in a data set.	**Medidas de Variabilidad**	Las estadísticas que ayudan a determinar la extensión de los números en una serie de datos.
Median	The middle number or the average of the two middle numbers in an ordered data set.	**Mediana**	El número medio o el promedio de los dos números medios en un conjunto de datos ordenados.
Minimum	The lowest point on a curve.	**Mínimo**	El punto más bajo en la curva.

minimum

mínimo

English		Español	
Mixed Number	The sum of a whole number and a fraction less than 1.	**Números Mixtos**	La suma de un número entero y una fracción menor que 1.
Mode	The number(s) or item(s) that occur most often in a data set.	**Moda**	El número(s) o artículo(s) que ocurre con más frecuencia en un conjunto de datos.
Motion Rate	A rate that compares distance to time.	**Índice de Movimiento**	Un índice que compara distancia por tiempo.
Multiple	The product of a number and nonzero whole number.	**Múltiplo**	El producto de un número y un número entero que no sea cero.

N

Negative Number	A number less than 0.	**Número Negativo**	Un número menor que 0.

Net	A two-dimensional pattern that folds to form a solid.	**Red**	Un patrón bidimensional que se dobla para formar un sólido.

Non-Linear Function	A function whose graph does not form a line.	**Ecuación No Lineal**	Una ecuación cuya gráfica no forma una recta.
Normal Distribution	A set of data values where the majority of the values are located in the middle of the data set and can be displayed by a bell-shaped curve.	**Distribución Normal**	Un conjunto de valores de datos donde la mayoría de los valores están localizados en el medio del conjunto de datos y pueden ser mostrados por una curva de forma de campana.
Numerical Data	Data collected in the form of numbers.	**Datos Numéricos**	Datos recopilados en la forma de números.
Numerical Expressions	An expression consisting of numbers and operations that represents a specific value.	**Expresiones Numéricas**	Una expresión que consta de números y operaciones que representa un valor específico.

O

Obtuse Angle	An angle that measures more than 90° but less than 180°.	**Ángulo Obtuso**	Un ángulo que mide más de 90° pero menos de 180°.

Opposites	Numbers that are the same distance from 0 on a number line but are on opposite sides of 0.	**Opuestos**	Números a la misma distancia del 0 en un recta numérica pero en lados opuestos del 0.
Order of Operations	The rules to follow when evaluating an expression with more than one operation.	**Orden de Operaciones**	Las reglas a seguir cuando se evalúa una expresión con más de una operación.
Ordered Pair	A pair of numbers used to locate a point on a coordinate plane (x, y).	**Par Ordenado**	Un par de números utilizados para localizar un punto en un plano de coordenadas (x, y).

Origin	The point where the *x*- and *y*-axis intersect on a coordinate plane (0, 0).	**Origen**	El punto donde el eje de la *x*- *y* el de la *y*- se cruzan en un plano de coordenadas (0,0).

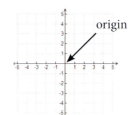

Outcome	One possible result from an experiment or a probability sample space.	**Resultado**	Un resultado posible de un experimento o un espacio de probabilidad de la muestra.
Outlier	An extreme value that varies greatly from the other values in a data set.	**Dato Aberrante**	Un valor extremo que varía mucho de los otros valores en un conjunto de datos.

P

Parabola	The graph of a quadratic function.	**Parábola**	La gráfica de una función cuadratica.

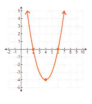
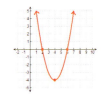

Parallel	Lines in the same plane that never intersect.	**Paralela**	Rectas en el mismo plano que nunca se intersecan.
Parallel Box-and-Whisker Plot	One box-and-whisker plot placed above another; often used to compare data sets.	**Diagrama Paralelo de Líneas y Bloques**	Un diagrama de líneas y bloques ubicado sobre otro para comparar conjuntos de datos.

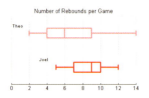
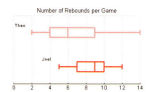

Parallelogram	A quadrilateral with both pairs of opposite sides parallel.	Paralelogramo	Un cuadrilátero con ambos pares de lados opuestos paralelos.

Parent Function	The simplest form of a particular type of function.	Función Principal	La forma más simple de un tipo particular de la función.
Parent Graph	The most basic graph of a function.	Gráfico Matriz	La gráfica más básica de una función.
Percent	A ratio that compares a number to 100.	Por Ciento	Una razón que compara un número con 100.
Percent of Change	The percent a quantity increases or decreases compared to the original amount.	Por Ciento de Cambio	El por ciento que una cantidad aumenta o disminuye comparado a la cantidad original.
Percent of Decrease	The percent of change when the new amount is less than the original amount.	Por Ciento de Disminución	El por ciento de cambio cuando la nueva cantidad es menos que la cantidad original.
Percent of Increase	The percent of change when the new amount is more than the original amount.	Por Ciento de Incremento	El por ciento de cambio cuando la nueva cantidad es más que la cantidad original.
Perfect Cube	A number whose cube root is an integer.	Cubo Perfecto	Un número cuyo raíz cúbica es un número entero.
Perfect Square	A number whose square root is an integer.	Cuadrado Perfecto	Un número cuyo raíz cuadrado es un número entero.
Perfect Square Trinomial	A trinomial that is the square of a binomial.	Trinomio Cuadrado Perfecto	Un trinomio que es el cuadrado de un binomio.
Perimeter	The distance around a figure.	Perímetro	La distancia alrededor de una figura.

Perpendicular	Two lines or segments that form a right angle.	**Perpendicular**	Dos rectas o segmentos que forman un ángulo recto.

Pi (π)	The ratio of the circumference of a circle to its diameter.	**Pi (π)**	La razón de la circunferencia de un círculo a su diámetro.
Pictograph	A graph that uses pictures to compare the amounts represented in a categorical data set.	**Gráfica Pictórica**	Una gráfica que utiliza dibujos para comparar las cantidades representadas en un conjunto de datos categóricos.

Pie Chart	A circle graph that shows information as sectors of a circle.	**Gráfico Circular**	Enseña la información como sectores de un círculo.

Polygon	A closed figure formed by three or more line segments.	**Polígono**	Una figura cerrada formada por tres o más segmentos de rectas.
Population	The entire group of people or objects one wants to gather information about.	**Población**	Todo el grupo de personas o los objetos a los que se quiere obtener información sobre.
Positive Number	A number greater than 0.	**Número Positivo**	Un número mayor que 0.
Power	An expression such as x^a which consists of two parts, the base (x) and the exponent (a).	**Potencia**	Una expresión como x^a que consiste de dos partes, la base (x) y el exponente (a).
Pre-image	The original figure prior to a transformation.	**Pre-imagen**	La figura original antes de una transformación.

Prime Factorization	When any composite number is written as the product of all its prime factors.	**Factorización Prima**	Cuando cualquier número compuesto es escrito como el producto de todos los factores primos.
Prime Number	A whole number larger than 1 that has only two possible factors, 1 and itself.	**Número Primo**	Un número entero mayor que 1 que tiene solo dos factores posibles, 1 y el mismo.
Prism	A solid formed by polygons with two congruent, parallel bases.	**Prisma**	Un sólido formado por polígonos con dos bases congruentes y paralelas.
Probability	The measure of how likely it is an event will occur.	**Probabilidad**	La medida de cuán probable un suceso puede ocurrir.
Product	The answer to a multiplication problem.	**Producto**	La respuesta a un problema de multiplicación.
Proper Fraction	A fraction with a numerator that is less than the denominator.	**Fracción Propia**	Una fracción con un numerador que es menos que el denominador.
Proportion	An equation stating two ratios are equivalent.	**Proporción**	Una ecuación que establece que dos razones son equivalentes.
Protractor	A tool used to measure angles.	**Transportador**	Una herramienta para medir ángulos.
Pyramid	A solid with a polygonal base and triangular sides that meet at a vertex.	**Pirámide**	Un sólido con una base poligonal y lados triangulares que se encuentran en un vértice.

Pythagorean Triple	A set of three positive integers (a, b, c) such that $a^2 + b^2 = c^2$.	Triple de Pitágoras	Un conjunto de tres enteros positivos (a, b, c) tal que $a^2 + b^2 = c^2$.

Q

Q-Points	Points that are created by the intersection of the quartiles for the x- and y-values of a two-variable data set.	Puntos Q	Puntos que son creados por la intersección de los cuartiles para los valores de la x- y la y- de un conjunto de datos de dos variables.
Quadrants	Four regions formed by the x and y axes on a coordinate plane.	Cuadrantes	Cuatro regiones formadas por el eje-x y el eje-y en un plano de coordenadas.

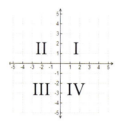

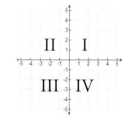

Quadratic Formula	A method which can be used to solve quadratic equations in the form $0 = ax^2 + bx + c$, where $a \neq 0$. $$x = \frac{-b \pm \sqrt{b^2 - 4ac}}{2a}$$	Fórmula Cuadrática	Un métado que puede usarse para resolver ecuaciones cuadraticas en la forma $0 = ax^2 + bx + c$, donde $a \neq 0$. $$x = \frac{-b \pm \sqrt{b^2 - 4ac}}{2a}$$
Quadratic Function	Any function in the family with the parent function of $f(x) = x^2$.	Función Cuadrática	Cualquier otra función en la familia con la función principal de $f(x) = x^2$.
Quadrilateral	A polygon with four sides.	Cuadrilateral	Un polígono con cuatro lados.
Quotient	The answer to a division problem.	Cociente	La solución a un problema de división.

R

Radius	The distance from the center of a circle to any point on the circle.	Radio	La distancia desde el centro de un círculo a cualquier punto en el círculo.

English		Spanish	
Random Sample	A group from a population created when each member of the population is equally likely to be chosen.	Muestra Aleatoria	Un grupo de una población creada cuando cada miembro de la población tiene la misma probabilidad de ser elegido.
Range (of a data set)	The difference between the maximum and minimum values in a data set.	Rango	La diferencia entre los valores máximo y mínimo en un conjunto de datos.
Range (of a function)	The set of output values of a function.	Rango (de una función)	El conjunto de valores salidos de la función.
Rate	A ratio of two numbers that have different units.	Índice	Una proporción de dos números con diferentes unidades.
Rate Conversion	A process of changing at least one unit of measurement in a rate to a different unit of measurement.	Conversión de Índice	Un proceso de cambiar por lo menos una unidad de medición en un índice a una diferente unidad de medición.
Rate of Change	The change in y-values over the change in x-values on a linear graph.	Índice de Cambio	El cambio en los valores de y sobre el cambio en los valores de x en una gráfica lineal.
Ratio	A comparison of two numbers using division. $a : b \quad \frac{a}{b} \quad a \text{ to } b$	Razón	Una comparación de dos números utilizando división. $a : b \quad \frac{a}{b} \quad a \text{ a } b$
Rational Number	A number that can be expressed as a fraction of two integers.	Número Racional	Un número que puede ser expresado como una fracción de dos enteros.
Ray	A part of a line that has one endpoint and extends forever in one direction.	Rayo	Una parte de una recta que tiene un punto final y se extiende eternamente en una dirección.
Real Numbers	The set of numbers that includes all rational and irrational numbers.	Números Racionales	El conjunto de números que incluye todos los números racionales e irracionales.

Reciprocals	Two numbers whose product is 1.	**Recíprocos**	Dos números cuyo producto es 1.
Rectangle	A quadrilateral with four right angles.	**Rectángulo**	Un cuadrilátero con cuatro ángulos rectos.

Recursive Routine	A routine described by stating the start value and the operation performed to get the following terms.	**Rutina Recursiva**	Una rutina descrita al exponer el valor del comienzo y la operación realizada para conseguir los términos siguientes.
Recursive Sequence	An ordered list of numbers created by a first term and a repeated operation.	**Secuencia Recursiva**	Una lista de números ordenados creada por un primer término y una operación repetida.
Reduction	A dilation that creates an image smaller than its pre-image.	**Reducción**	Una dilatación que crea una imagen más pequeña que su pre-imagen.
Reflection	A transformation in which a mirror image is produced by flipping a figure over a line.	**Reflexión**	Una transformación en el que se produce una imagen reflejada volteando una figura sobre una línea.

Relative Frequency	The ratio of the observed frequency to the total number of frequencies in an experiment or survey.	**Frecuencia Relativa**	La proporción de la frecuencia observada para el número total de frecuencias en un experimento o estudio.
Remainder	A number that is left over when a division problem is completed.	**Remanente**	Un número que queda cuando un problema de división se ha completado.
Repeating Decimal	A decimal that has one or more digits that repeat forever.	**Decimal Repetitivo**	Un decimal que tiene uno o más dígitos que se repiten eternamente.

Representative Sample	A group from a population that accurately represents the entire population.	**Muestra Representativa**	Un grupo de una población que representa con precisión toda la población.

Rhombus	A quadrilateral with four sides equal in measure.	**Rombo**	Un cuadrilátero con cuatro lados iguales en la medida.

Right Angle	An angle that measures 90°.	**Ángulo Recto**	Un ángulo que mide 90°.

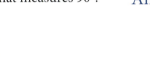

Roots	The x-intercepts of a quadratic function.	**Raíces**	Las intersecciones-x de una función cuadratica.

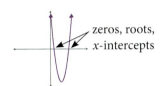

zeros, roots, x-intercepts

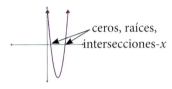

ceros, raíces, intersecciones-x

Rotation	A transformation which turns a point or figure about a fixed point, often the origin.	**Rotación**	Una transformación que convierte un punto a una figura sobre un punto fijo.

S

Sales Tax	An amount added to the cost of an item. The amount added is a percent of the original amount as determined by a state, county or city.	**Impuesto sobre las Ventas**	Una cantidad añadida al costo de un artículo. La cantidad añadida es un por ciento de la cantidad original determinado por el estado, condado o ciudad.

| Same-Side Interior Angles | Two angles that are on the inside of two lines and are on the same side of a transversal. | Ángulos Interiores del Mismo Lado | Dos ángulos que están en el interior de dos rectas y están en el mismo lado de una transversal. |

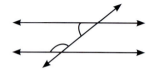 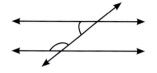

| Sample | A group from a population that is used to make conclusions about the entire population. | Muestra | Un grupo de una población que se utiliza para sacar conclusiones sobre toda la población. |

| Sample Space | The set of all possible outcomes. | Muestra de Espacio | El conjunto de todos los resultados posibles. |

| Scale | The ratio of a length on a map or model to the actual object. | Escala | La razón de una longitud en un mapa o modelo al objeto verdadero. |

| Scale Factor | The ratio of corresponding sides in two similar figures. | Factor de Escala | La razón de los lados correspondientes en dos figuras similares. |

| Scalene Triangle | A triangle that has no congruent sides. | Triángulo Escaleno | Un triángulo sin lados congruentes. |

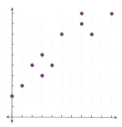 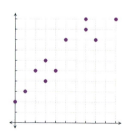

| Scatter Plot | A set of ordered pairs graphed on a coordinate plane. | Diagrama de Dispersión | Un conjunto de pares ordenados graficados en un plano de coordenadas. |

| Scientific Notation | Scientific notation is an exponential expression using a power of 10 where $1 \leq N < 10$ and P is an integer. $N \times 10^P$ | La Notación Científica | Notación científica es una expresión exponencial con una potencia de 10, donde $1 \leq N < 10$ y P es un número entero. $N \times 10^P$ |

Sector	A portion of a circle enclosed by two radii.	Sector	Una porción de un circulo encerado por dos radios.

Sequence	An ordered list of numbers.	Sucesión	Una lista de números ordenados.

Similar Figures	Two figures that have the exact same shape, but not necessarily the exact same size.	Figuras Similares	Dos figuras que tienen exactamente la misma forma, pero no necesariamente el mismo tamaño exacto.

Similar Solids	Solids that have the same shape and all corresponding dimensions are proportional.	Sólidos Similares	Sólidos con la misma forma y todas sus dimensiones correspondientes son proporcionales.

Simplest Form	A fraction whose numerator and denominator's only common factor is 1.	Expresión Simple	Una fracción cuyo único factor común del numerador y del denominador es 1.

Simplify an Expression	To rewrite an expression without parentheses and combine all like terms.	Simplificar una Expresión	Reescribir una expresión sin paréntesis y combinar todos los términos iguales.

Simulation	An experiment used to model a situation.	Simulación	Un experimento utilizado para modelar una situación.

Single-Variable Data	A data set with only one type of data.	Datos de una Variable	Un conjunto de datos con tan solo un tipo de datos.

Sketch	To make a figure free hand without the use of measurement tools.	Esbozo	Hacer una figura a mano libre sin utilizar herramientas de medidas.

Skewed Left	A plot or graph with a longer tail on the left-hand side.	Torcido a la Izquierda	Un gráfico con una cola al lado izquierdo.

Skewed Right	A plot or graph with a longer tail on the right-hand side.	**Torcido a la Derecha**	Un gráfico con una cola al lado derecho.
Slant Height	The height of a lateral face of a pyramid or cone.	**Altura Sesgada**	La altura de un cara lateral de una pirámide o cono.

Slope	The ratio of the vertical change to the horizontal change in a linear graph.	**Pendiente**	La razón del cambio vertical al cambio horizontal en una gráfica lineal.
Slope Triangle	A right triangle formed where one leg represents the vertical rise and the other leg is the horizontal run in a linear graph.	**Triángulo de Pendiente**	Un triángulo rectángulo formado donde una cateto representa el ascenso y la otra es una carrera horizontal en una gráfica lineal.

Slope-Intercept Form	A linear equation written in the form $y = mx + b$.	**Forma de las Intersecciones con la Pendiente**	Una ecuación lineal escrita en la forma $y = mx + b$.
Solid	A three-dimensional figure that encloses a part of space.	**Sólido**	Una figura tridimensional que encierra una parte del espacio.
Solution	Any value or values that makes an equation true.	**Solución**	Cualquier valor o valores que hacen una ecuación verdadera.
Solution of a System of Linear Equations	The ordered pair that satisfies both linear equations in the system.	**Solución de un Sistema de Ecuaciones Lineales**	El par ordenado que satisface ambas ecuaciones lineales en el sistema.

| Sphere | A solid formed by a set of points in space that are the same distance from a center point. | Esfera | Un sólido formado por un conjunto de puntos en el espacio que están a la misma distancia de un punto central. |

| Square | A quadrilateral with four right angles and four congruent sides. | Cuadrado | Un cuadrilátero con cuatro ángulos rectos y cuatro lados congruente. |

| Square Root | One of the two equal factors of a number. $$3 \cdot 3 = 9 \qquad 3 = \sqrt{9}$$ | Raíz Cuadrada | Uno de los dos factores iguales de un número. $$3 \cdot 3 = 9 \qquad 3 = \sqrt{9}$$ |

| Squared | A term raised to the power of 2. | Cuadrado | Un término elevado a la potencia de 2. |

| Start Value | The output value that is paired with an input value of 0 in an input-output table. | Valor de Comienzo | El valor de salida que es aparejado con un valor de entrada de 0 en una tabla de entradas y salidas. |

| Statistics | The process of collecting, displaying and analyzing a set of data. | Estadísticas | El proceso de recopilar, exponer y analizar un conjunto de datos. |

| Stem-and-Leaf Plot | A plot which uses the digits of the data values to show the shape and distribution of the data set. | Gráfica de Tallo y Hoja | Un diagrama que utiliza los dígitos de los valores de datos para mostrar la forma y la distribución del conjunto de datos. |

| Straight Angle | An angle that measures 180°.  | Ángulo Recto | Un ángulo que mide 180°.  |

Straight Edge	A ruler-like tool with no markings.	Borde Recto	Un gobernante como herramienta sin marcas.
Substitution Method	A method for solving a system of linear equations.	Método de Substitución	Un método para resolver un sistema de ecuaciones lineales.
Supplementary Angles	Two angles whose sum is 180°.	Ángulos Suplementarios	Dos ángulos cuya suma es 180°.
Surface Area	The sum of the areas of all the surfaces on a solid.	Área de la Superficie	La suma de las áreas de todas las superficies en un sólido.
System of Linear Equations	Two or more linear equations.	Sistema de Ecuaciones Lineales	Dos o más ecuaciones lineales.

T

Term	A number or the product of a number and a variable in an algebraic expression; a number in a sequence.	Término	Un número o el producto de un número y una variable en una expresión algebraica; un número en una sucesión.
Terminating Decimal	A decimal that stops.	Decimal Terminado	Un decimal que para.
Theorem	A relationship in mathematics that has been proven.	Teorema	Una relación en las matemáticas que ha sido probada.
Theoretical Probability	The ratio of favorable outcomes to the number of possible outcomes.	Probabilidad Teórica	La proporción de resultados favorables a la cantidad de resultados posibles.
Third Quartile (Q3)	The median of the upper half of a data set.	Tercer Cuartil (Q3)	Mediana de la parte superior de un conjunto de datos.
Tick Marks	Equally divided spaces marked with a small line between every inch or centimeter on a ruler.	Marcas de Graduación	Espacios divididos igualmente marcados con una línea pequeña entre cada pulgada o centímetro en una regla.
Transformation	The movement of a figure on a graph so that it changes size or position.	Transformación	El movimiento de una figura en un gráfico de modo que cambia el tamaño o posición

| Translation | A transformation in which a figure is shifted up, down, left or right. | Traducción | Una transformación donde la figura se mudo arriba, abajo, a la izquierda o a la derecha. |

 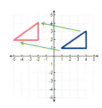

| Transversal | A line that intersects two or more lines in the same plane. | Transversal | Una recta que interseca dos o más rectas en el mismo plano. |

| Trapezoid | A quadrilateral with exactly one pair of parallel sides. | Trapezoide | Un cuadrilateral con exactamente un par de lados paralelos. |

 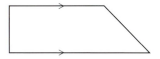

| Tree Diagram | A display that organizes information to determine possible outcomes. | Diagrama de Árbol | Una pantalla que organiza la información para determinar los posibles resultados. |

 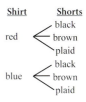

| Trial | A single act of performing an experiment. | Prueba | Un solo intento de realizar un experimento. |

| Trinomial | An expression with three terms (i.e. $x^2 - 3x + 4$). | Trinomio | Una expreción que tiene tres terminos (es decir: $x^2 - 3x + 4$). |

| Two-Step Equation | An equation that has two different operations. | Ecuación de Dos Pasos | Una ecuación que tiene dos operaciones diferentes. |

| Two-Variable Data | A data set where two groups of numbers are looked at simultaneously. | Datos de dos Variables | Un conjunto de datos dónde dos grupos de números se observan simultáneamente. |

Two-Way Frequency Table	A table that shows how many times a value occurs for a pair of categorical data.	Tabla de Frecuencia Bidireccional	Una tabla que muestra cuántas veces aparece un valor de un par de datos categóricos.

		Walk	
		Yes	No
Dog Owners	Yes	15	20
	No	25	20

		Paseo	
		Si	No
Perro Propietario	Si	15	20
	No	25	20

U-V-W

Unit Rate	A rate with a denominator of 1.	Índice de Unidad	Un índice con un denominador de 1.

Univariate Data	Data that describes one variable (i.e., scores on a test).	Data Univariados	Datos que describen una variable (es decir: puntajes en una prueba).

Variable	A symbol that represents one or more numbers.	Variable	Un símbolo que representa uno o más números.

Vertex	The minimum or maximum point on a parabola.	Vértice	El mínimo o máximo punto en una parábola.

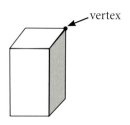

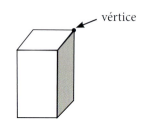

Vertex of a Solid	The point where three or more edges meet.	Vértice de un Sólido	El punto donde tres o más bordes se encuentran.

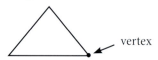

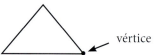

Vertex of a Triangle	A point where two sides of a triangle meet.	Vértice de un Triángulo	Un punto donde dos lados de un triángulo se encuentran.

Vertex of an Angle	The common endpoint of the two rays that form an angle. 	**Vértice de un Ángulo**	El punto final en común de los dos rayos que forma un ángulo. 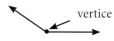
Vertex Form	A quadratic function is in vertex form when written $f(x) = a(x - h)^2 + k$ where $a \neq 0$.	**Forma De Vértice**	Una función cuadrática es en forma general cuándo escrito $f(x) = a(x - h)^2 + k$ donde $a \neq 0$.
Vertical Angles	Non-adjacent angles with a common vertex formed by two intersecting lines. 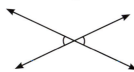	**Ángulos Verticales**	Ángulos no adyacentes con un vértice en común formado por dos rectas intersecantes. 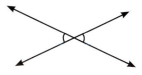
Vertical Line Test	A test used to determine if a graph represents a function by checking to see if a vertical line passes through no more than one point of the graph of a relation.	**Examen Vertical De Línia**	Un examen para determinar si una gráfica representa una función. Es utilizada para ver si una línia vertical que pasa a través de no más de un punto de la gráfica de una relación.
Volume	The number of cubic units needed to fill a three-dimensional figure.	**Volumen**	La cantidad de unidades cúbicas necesitadas para llenar un sólido.

X-Y-Z

x-Axis	The horizontal number line on a coordinate plane. 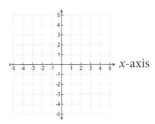	Eje-*x*, Eje de la *x*	La recta numérica horizontal en un plano de coordenadas. 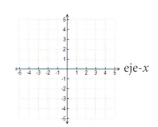

y-Axis	The vertical number line on a coordinate plane.	Eje-*y*, Eje de la *y*	La recta numérica vertical en un plano de coordenadas.

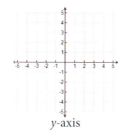

y-axis

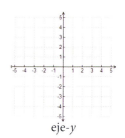

eje-y

y-Intercept	The point where a graph intersects the *y*-axis.	Intersección *y*	El punto donde una gráfica interseca el eje-*y*.

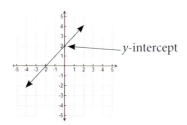

y-intercept

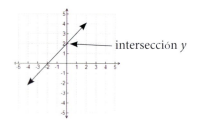

intersección y

Zero Pair	One positive integer chip paired with one negative integer chip.	Par Cero	Un chip entero positivo emparejado con un chip entero negativo.

 + = 0

1 + (−1) = 0

 + = 0

1 + (−1) = 0

Zero Product Property	If a product of two factors is equal to zero, then one or both of the factors must be zero.	Propiedad De Producto Cero	Si un producto de dos factores es iqual a cero, uno o ambos de los factores debe ser cero.
Zeros	The *x*-intercepts of a quadratic function.	Ceros	Las intersecciones-*x* de una función cuadratica.

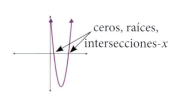

zeros, roots, *x*-intercepts

ceros, raíces, intersecciones-*x*

SELECTED ANSWERS

BLOCK 1

Lesson 1.1

1. 1.12 **3.** 2.39 **5.** 0.3 **7.** 2.3 **9. a)** thousandths **b)** ones **c)** tenths **d)** hundredths **11.** eight thousandths **13.** three ones **15.** 99.2 **17.** 52.06 **19.** 0.223 **21.** thirty-five and eighty-seven hundredths **23.** five hundred forty-nine thousandths **25.** sixty-eight and four tenths

Lesson 1.2

1. 4 **3.** 7 **5.** 18 **7.** approx. 2 hours **9.** 34.9 **11.** 71.3 **13.** 3.9 **15.** $2.40 **17.** 45.21 **19.** 321.24 **21.** 3.90 **23.** 7.909 **25.** 201.109 **27.** 5.001 **29.** 1.34 **31.** 0.27 **33.** seventy and seven hundredths **35.** 8 hundredths **37.** 1 thousandth

Lesson 1.3

1. 10 **3.** 5 *cm* **5.** 14.1 *cm* **7.** 9 *cm* **9.** 14.1 *cm* **11.** 4.1 *cm* **13. a)** Answers may vary **b)** 6.6 *cm* **c)** Answers may vary **15.** 7.5 *cm* **17.** 16.5 *cm* **19. a)** Answers may vary **b)** 7.5 *cm* **c)** Answers may vary **21.** draw line 3.4 *cm* **23.** draw line 5.7 *cm* **25.** draw line 0.9 *cm* **27.** 35 *cm* **29.** 15.7 **31.** 321.4 **33.** 215.17

Lesson 1.4

1. < **3.** > **5.** > **7.** > **9.** < **11.** 17.8, 17.801, 17.81, 17.851 **13.** C. 6.9 **15.** D. 8.8 **17.** Idaho **19.** 48.9 seconds **21.** 51.5 seconds **23.** 50.5 and 50.98 seconds **25.** 48.9, 49.09, 49.25, 49.45, 49.61, 49.76, 50.5, 50.98, 51.5 **27. a)** Tuesday **b)** Wednesday **c)** 68.08, 68.75, 68.8 **29.** 32.63 **31.** 2.17 **33.** 6.4

Lesson 1.5

1. 8 **3.** 13 **5.** 7 **7.** 26 **9.** 22 **11.** About $15.00 **13.** 16 **15.** 27 **17.** 216 **19.** About $209.00 **21.** 5 **23.** 3 **25.** 1 **27.** approx. 10 books **29.** = **31.** 32.34, 32.4, 32.43, 32.48 **33.** 11.02, 11.022, 11.2, 11.22 **35.** See student work. **37.** See student work. **39.** See student work.

Lesson 1.6

1. 5.5 **3.** 13.206 **5.** 7.105 **7.** $20.13 **9.** 2.56 **11.** 2.86 **13.** 3.649 **15.** 8.31 inches **17.** 9.36 more inches **19.** 14.09 inches **21.** 10.92 inches **23.** $204.37 **25.** Answers may vary **27.** 4.68 **29.** 4.381 **31.** 36.8 *cm*

Block 1 Review

1. 1.34 **3.** 0.28 **5.** tenths **7.** 2.7 **9.** 0.28 **11.** nine and fifteen hundredths **13.** 52.99 **15.** 6.359 **17.** 9.0 **19.** 93.01 **21.** 60 **23.** $1.40 **25.** 9.3 *cm* **27.** 1.7 *cm* **29.** 2 *cm* **31. a)** Answers may vary **b)** 7.9 *cm* **c)** Answers may vary **33.** < **35.** = **37.** > **39.** 0.8, 0.842, 0.88, 0.884 **41.** 13 **43.** 28 **45.** 29 **47.** Approx. 13 pounds **49.** 9 **51.** About 29 miles **53.** 1.299 **55.** 16.048 **57.** 13.064 **59.** 1.96 **61.** 8.15 ounces

BLOCK 2

Lesson 2.1

1. 102 **3.** 238 **5.** 215 **7.** 360 minutes **9.** 448 **11.** 2,106 **13.** 2,535 **15.** 1,288 beads **17.** 192 ounces **19.** 4,278 **21.** 12,519 **23.** 856 *m* **25.** 7.2 *cm*, 7.8 *cm*, 8.3 *cm*, 8.6 *cm* **27.** 0.3 *cm*, 0.33 *cm*, 3 *cm*, 3.3 *cm* **29.** 1.8 *km*

Lesson 2.2

1. 12.3 **3.** $15.19 **5.** $24.45 **7.** $5.85 **9.** 23.68 **11.** 34.44 **13.** 19.448 **15.** 0.45 **17.** $0.28 **19.** $2.76 **21.** 0.0525 **23.** 0.05, 0.09, 0.1 **25.** 1.089, 1.31, 1.4 **27.** 13 **29.** 6 **31.** 17

Lesson 2.3

1. 15 R1 **3.** 16 R1 **5.** 25 R2 **7.** 14 tables **9.** 19 people per row, 3 left over **11.** $34\frac{1}{6}$ **13.** $175\frac{1}{2}$ **15.** $81\frac{5}{7}$ **17.** $108\frac{3}{7}$ pounds of fruit **19.** 36 R4 and $36\frac{4}{8}$ or $36\frac{1}{2}$ **21.** 196 R2 and $196\frac{2}{5}$ **23.** 16 R3 and $16\frac{3}{4}$ **25.** 133 R4 and $133\frac{4}{6}$ or $133\frac{2}{3}$ **27.** 62 **29.** 13.94 **31.** 16.744 **33.** 0.0455 **35.** 17.411

Lesson 2.4

1. 28 **3.** 30 R5 **5.** 31 R3 **7.** $20\frac{10}{20}$ or $20\frac{1}{2}$ **9.** $21\frac{28}{42}$ or $21\frac{2}{3}$ **11.** $21\frac{14}{35}$ or $21\frac{2}{5}$ **13.** $16\frac{9}{18}$ or $16\frac{1}{2}$ gallons each day **15. a)** 52 boxes **b)** 24 boxes, 3 bars left over **c)** 36 boxes, 6 bars left over **d)** 45 boxes, 6 bars left over **17.** 216 R4 and $216\frac{4}{13}$ **19.** 201 R7 and $201\frac{7}{15}$ **21.** 292 R4 and $292\frac{4}{15}$ **23.** 5.2 **25.** 8.206 **27.** 26 tables, 7 plates left over

Lesson 2.5

1. 1.9 **3.** 7.3 **5.** 2.8 **7.** $1.49 **9.** $0.12 **11.** 0.0875 **13.** 0.098 **15.** 1.15 **17.** $1.50 **19.** 1.24 **21.** 3.97 **23.** 5.37 **25. a)** 16 **b)** 936 **27. a)** 207 **b)** 255

Lesson 2.6

1. 48 ÷ 8 **3.** 1240 ÷ 34 **5.** 182.9 ÷ 31 **7.** 9 **9.** 70 **11.** 35 **13.** 65.6 **15.** 1.7 **17.** 3.45 times heavier **19.** 4.4 **21.** 4 **23.** 10 **25.** $1.28/pound **27.** 1.6 times greater **29.** 42.84 **31.** 0.20 **33.** 7.25

Block 2 Review

1. 280 **3.** 574 **5.** 777 **7.** 544 **9.** 2,613 **11.** 768 pages
13. 18.63 **15.** 8.84 **17.** 37.2422 **19.** 0.3471 **21.** $2.63
23. 25 **25.** 11 R5 **27.** 59 R4 **29. a)** Raul **b)** $\frac{3}{873}$ in quotient should be R3 or $\frac{3}{6} = \frac{1}{2}$ **31.** 31 R20 **33.** 120 R3 **35.** 153 R2
37. 16 flowers/arrangement, 8 flowers left **39.** 4.95 **41.** 6.5
43. 0.06 **45.** 3.22 **47.** 5.525 **49.** 22.33 **51.** 6.98 **53.** 1.66
55. 2.6 times greater

BLOCK 3

Lesson 3.1

1. 1, 2, 4 composite **3.** 1, 2, 4, 8 composite **5.** 1, 2, 7, 14 composite **7.** 1, 29 prime **9.** 1, 3, 9, 27 composite
11. 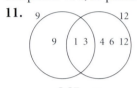 **13.**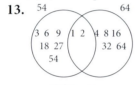

GCF = 3   GCF = 2

15. 9: 1, ③, 9   **17.** 48: 1, 2, 3, 4, 6, 8, 12, ⑯, 24, 48
    6: 1, 2, ③, 6   32: 1, 2, 4, 8, ⑯, 32
                  64: 1, 2, 4, 8, ⑯, 32, 64
19. 24: 1, ②, 3, 4, 6, 8, 12, 24
    40: 1, ②, 4, 5, 8, 10, 20, 40
    54: 1, ②, 3, 6, 9, 18, 27, 54
21. Prime factor trees may vary: 18 = **3** × 3 × **2**
    24 = **3** × **2** × 2 × 2 GCF = 3 × 2 = 6
23. Prime factor trees may vary: 84 = **2** × **2** × 3 × **7**
    56 = **2** × **2** × 2 × **7** GCF = 2 × 2 × 7 = 28
25. 18 candy bars **27.** 1.62 **29.** 20.60 **31.** 12.46 **33.** 3,301
35. 171.15

Lesson 3.2

1. a) b) c) $\frac{1}{3}$ ▮ $\frac{2}{6}$ **d)** Yes, $\frac{1}{3}$ is the same as $\frac{2}{6}$
3. $\frac{2}{5}$ $\frac{3}{6}$ **5.** No, they are not equivalent. The model for $\frac{5}{8}$ has less colored in than the model for $\frac{3}{4}$ **7.** 4 **9.** 30 **11.** 3 **13.** 6 **15.** 4 **17.** 9
19. Answers may vary. $\frac{2}{6}, \frac{3}{9}$ **21.** Answers may vary. $\frac{2}{5}, \frac{8}{20}$
23. Answers may vary. $\frac{2}{3}, \frac{12}{18}$ **25.** Yes. $\frac{3}{9}$ is equivalent to $\frac{1}{3}$
27. 1, 11 prime **29.** 1, 3, 13, 39 composite **31.** 4

Lesson 3.3

1. simplest form **3.** $\frac{4}{5}$ **5.** simplest form **7.** $\frac{1}{3}$ **9.** simplest form
11. $\frac{2}{7}$ **13.** $\frac{3}{5}$ **15.** $\frac{1}{3}$ **17.** $\frac{3}{8}$ **19.** $\frac{2}{5}$ **21.** $\frac{8}{9}$ **23.** $\frac{12}{24} = \frac{1}{2}$ and $\frac{16}{36} = \frac{4}{9}$
No, they are not equivalent **25.** Answers may vary.
27. $\frac{3}{8}$ pound **29.** $\frac{36}{126} = \frac{2}{7}$ of the students **31.** Answers may vary. $\frac{2}{6}, \frac{3}{9}$ **33.** Answers may vary. $\frac{4}{5}, \frac{8}{10}$ **35.** 2

Lesson 3.4

1. 2, 4, 6, 8, 10 **3.** 14, 28, 42, 56, 70 **5.** 20 **7.** 30 **9.** 24
11. 70 **13.** 42 **15.** 60 **17.** 35 **19.** 14 **21.** 24 **23.** Friday
25. Tuesday **27.** 15 **29.** 6 **31.** 30 **33.** 5

Lesson 3.5

1. $\frac{2}{3}$ ⓐ$\frac{4}{5}$ **3.** < **5.** = **7.** > **9.** < **11.** > **13.** $\frac{1}{5}, \frac{1}{3}, \frac{2}{5}$ **15.** $\frac{3}{10}, \frac{2}{5}, \frac{7}{10}$
17. $\frac{1}{3}, \frac{11}{15}, \frac{4}{5}$ **19.** Answers may vary $\frac{7}{15}$ **21.** Answers may vary. $\frac{1}{3}$ **23.** Answers may vary. $\frac{1}{2}$ **25.** Worker bees
27. Yellow **29.** 60 **31.** 28 **33.** 12 **35.** $\frac{1}{10}$ **37.** simplest form

Lesson 3.6

1. $\frac{7}{5}$ and $1\frac{2}{5}$ **3.** $\frac{7}{2}$ and $3\frac{1}{2}$ **5.** $2\frac{1}{3}$ **7.** $3\frac{1}{3}$ **9.** $4\frac{4}{7}$ **11.** $3\frac{1}{3}$ **13.** $6\frac{18}{25}$ feet
15. $\frac{14}{3}$ **17.** $\frac{18}{7}$ **19.** $\frac{41}{8}$ **21.** $\frac{57}{9}$ **23.** $\frac{54}{15}$ **25.** $3\frac{1}{4}, \frac{7}{2}, \frac{9}{2}$ **27.** $1\frac{1}{8}, \frac{3}{2}, \frac{7}{4}$
29. $10\frac{5}{8}$ inches **31.** 17.6958 **33.** 162.18 **35.** > **37.** 18 **39.** 5

Lesson 3.7

1. 16 **3.** 3 *in* **5.** 10 *in* **7.** 11 *in* **9.** $4\frac{1}{4}$ *in* **11.** $6\frac{1}{4}$ *in* **13.** $3\frac{1}{4}$ *in*
15. $4\frac{3}{4}$ *in* **17.** draw line $\frac{3}{8}$ *in* **19.** draw line $2\frac{1}{2}$ *in*
21. draw line 4 *in* **23.** $\frac{1}{2}$ *in* **25.** red and green
27. a) Antonio's **b)** Answers may vary. **29.** $\frac{2}{5}, \frac{7}{10}, \frac{4}{5}$ **31.** $1\frac{8}{9}$
33. $2\frac{1}{4}$

Block 3 Review

1. 1, 5 prime **3.** 1, 3, 7, 21 composite **5.** 7 **7.** 9 **9.** 8
11. Not equivalent ▮▮▮▮▮ ▮▮ ▯▯▯▯
13. 8 **15.** 49 **17.** 9 **19.** Answers may vary. $\frac{3}{5}$ and $\frac{12}{20}$
21. $\frac{3}{4}$ **23.** $\frac{4}{5}$ **25.** $\frac{8}{27}$ **27.** $\frac{11}{100}$ of a ton **29.** $\frac{2}{5}$ and $\frac{12}{25}$ - Not Equivalent **31.** 15, 30, 45, 60, 75 **33.** 40 **35.** 96 **37.** 24
39. 30 days **41.** > **43.** $\frac{3}{10}, \frac{2}{5}, \frac{3}{5}$ **45.** $\frac{2}{5}, \frac{3}{7}, \frac{4}{5}$ **47.** Monday **49.** $4\frac{3}{5}$
51. $\frac{23}{6}$ **53.** $\frac{89}{12}$ **55.** $1\frac{1}{3}, \frac{5}{3}, \frac{11}{6}$ **57.** $5\frac{1}{2}$ *in* **59.** 4 *in* **61.** $3\frac{1}{2}$ *in*
63. draw line $5\frac{1}{8}$ *in* **65.** draw line $\frac{7}{16}$ *in*

BLOCK 4

Lesson 4.1

1. $\frac{1}{2}$ **3.** 0 **5.** 2 **7.** 1 **9.** $\frac{1}{2}$ **11.** about $\frac{1}{2}$ of a pie **13.** 8 **15.** 16
17. 25 **19.** 3 **21.** 5 **23.** about 41 miles **25.** approx. 14 feet
27. $\frac{11}{3}$ **29.** 5, 10, 15, 20, 25 **31.** 8, 16, 24, 32, 40

Lesson 4.2

1. $\frac{1}{2}$ **3.** $\frac{3}{4}$ **5.** $1\frac{1}{3}$ **7.** $\frac{1}{4}$ **9.** $\frac{6}{11}$ **11.** $\frac{1}{5}$ mile **13.** $1\frac{1}{12}$ **15.** $\frac{5}{6}$ **17.** $\frac{1}{6}$
19. $\frac{2}{9}$ **21.** $\frac{1}{12}$ meter **23. a)** $\frac{3}{4}$ of her students **b)** $\frac{1}{4}$ of her students **25.** draw line $2\frac{1}{8}$ *in* **27.** draw line $1\frac{1}{2}$ *in* **29.** 1

Lesson 4.3

1. 9 **3.** $3\frac{3}{4}$ **5.** $5\frac{7}{10}$ **7.** $5\frac{11}{12}$ **9.** $4\frac{7}{12}$ **11.** $5\frac{1}{3}$ inches **13.** $1\frac{1}{2}$ **15.** $3\frac{1}{2}$ **17.** $1\frac{11}{15}$ **19.** $2\frac{19}{30}$ **21.** $\frac{2}{3}$ **23.** $\frac{3}{4}$ more hours **25.** $3\frac{1}{5}$ more tons **27.** $\frac{19}{20}$ **29.** $\frac{3}{10}$ **31.** $\frac{43}{60}$

Lesson 4.4

1. Answers may vary. **3.** $7\frac{2}{5}$ **5.** $8\frac{1}{2}$ **7.** $12\frac{1}{4}$ **9.** $1\frac{7}{10}$ **11.** $\frac{5}{6}$ **13.** $3\frac{3}{8}$ **15.** $4\frac{1}{2}$ inches **17.** $5\frac{3}{10}$ miles **19.** $1\frac{2}{3}$ minutes **21.** $1\frac{3}{8}$ feet **23.** $81.48

Lesson 4.5

1. $1\frac{1}{2}$ inches **3.** 6 inches **5.** $6\frac{1}{4}$ inches **7.** Answers may vary. **9.** $2\frac{3}{4}$ inches **11.** 5 inches **13.** 7 inches **15.** $6\frac{7}{8}$ inches **17.** $6\frac{1}{2}$ inches **19.** $111\frac{3}{4}$ yards **21.** $5\frac{1}{3}$ **23.** $6\frac{1}{12}$ **25.** $10\frac{19}{24}$ **27.** $9\frac{1}{3}$

Block 4 Review

1. $\frac{1}{2}$ **3.** $\frac{1}{2}$ **5.** 1 **7.** 5 **9.** 2 **11.** 14 **13.** $\frac{5}{8}$ **15.** $\frac{2}{3}$ **17.** $1\frac{1}{10}$ **19.** $\frac{1}{9}$ **21.** $\frac{11}{18}$ **23.** $\frac{5}{24}$ yard taller **25.** $4\frac{1}{4}$ **27.** $3\frac{1}{4}$ **29.** $7\frac{1}{2}$ **31.** $1\frac{17}{18}$ **33.** $6\frac{1}{12}$ quarts **35.** $5\frac{5}{7}$ **37.** $3\frac{9}{10}$ **39.** $1\frac{2}{3}$ **41.** $11\frac{1}{8}$ quarts **43.** 11 inches **45.** $12\frac{5}{8}$ inches **47.** $2\frac{3}{4}$ inches **49.** $281\frac{1}{2}$ feet

BLOCK 5

Lesson 5.1

1. 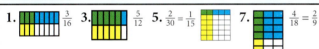 $\frac{3}{16}$ **3.** $\frac{5}{12}$ **5.** $\frac{2}{30} = \frac{1}{15}$ **7.** $\frac{4}{18} = \frac{2}{9}$

9. $\frac{4}{5} \times \frac{3}{4} = \frac{12}{20} = \frac{3}{5}$ **11.** $\frac{1}{6}$ cup **13.** $1\frac{7}{12}$ **15.** $1\frac{1}{2}$ **17.** $10\frac{11}{12}$

Lesson 5.2

1. $\frac{4}{15}$ **3.** $\frac{2}{15}$ **5.** $\frac{1}{5}$ **7.** $\frac{6}{35}$ **9.** $\frac{7}{12}$ **11.** $\frac{1}{4}$ of the routine **13.** $\frac{2}{5}$ **15.** $\frac{3}{14}$ **17.** $\frac{9}{13}$ **19.** $\frac{5}{12}$ **21.** $\frac{3}{10}$ of his shots **23.** $\frac{5}{8}$ hour **25.** 10 **27.** 6 inches **29.** 29 inches

Lesson 5.3

1. a) $\frac{3}{4} \div \frac{1}{4}$ or $\frac{6}{8} \div \frac{2}{8}$ **b) & c)** 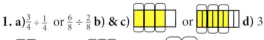 **d)** 3
3. 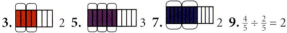 2 **5.** 3 **7.** 2 **9.** $\frac{4}{5} \div \frac{2}{5} = 2$
11. $\frac{6}{12} \div \frac{2}{12} = 3$ OR $\frac{1}{2} \div \frac{2}{12} = 3$ OR $\frac{6}{12} \div \frac{1}{6} = 3$ OR $\frac{1}{2} \div \frac{1}{6} = 3$
13. $\frac{6}{8} \div \frac{2}{8} = 3$ OR $\frac{3}{4} \div \frac{2}{8} = 3$ OR $\frac{6}{8} \div \frac{1}{4} = 3$ OR $\frac{3}{4} \div \frac{1}{4} = 3$
15. 4 times **17.** 8 slices **19.** $\frac{3}{8}$ **21.** 20 **23.** 30 **25.** $1\frac{19}{24}$

Lesson 5.4

1. 2 **3.** 4 **5.** 6 **7.** 4 **9.** 3 **11. a)** 2 **b)** 2 **c)** 4 **d)** 6 **13.** $1\frac{3}{7}$ **15.** $1\frac{2}{3}$ **17.** $1\frac{3}{7}$ **19.** $1\frac{1}{2}$ **21.** $5\frac{1}{3}$ sections **23.** $\frac{6}{8} \div \frac{2}{8} = 3$ OR $\frac{3}{4} \div \frac{2}{8} = 3$ OR $\frac{6}{8} \div \frac{1}{4} = 3$ OR $\frac{3}{4} \div \frac{1}{4} = 3$ **25.** $\frac{7}{8}$ **27.** $5\frac{11}{15}$

Lesson 5.5

1. a) Answers may vary. **b)** Answers may vary. **3.** 11 or 12 **5.** 12 **7.** 6 **9.** about 15 fly balls **11.** 18 **13.** 28 **15.** 14 **17.** About 30 cups **19.** 2 **21.** 9 **23.** 6 **25.** Approx. 7 days **27.** $\frac{3}{8}$ **29.** $\frac{7}{20}$ **31.** 2

Lesson 5.6

1. 4 **3.** 15 **5.** $5\frac{1}{3}$ **7.** 10 medals **9.** $19\frac{1}{5}$ windows **11.** 10 **13.** 15 **15.** 12 **17.** 15 people **19.** $22\frac{1}{2}$ trips **21.** $5\frac{1}{2}$ inches **23.** 5 inches **25.** 6 **27.** 5 **29.** 10

Lesson 5.7

1. $4\frac{1}{4}$ **3.** 11 **5.** $1\frac{1}{8}$ **7.** 4 **9.** $9\frac{1}{6}$ **11.** $7\frac{7}{8}$ cups flour **13.** 4 **15.** $9\frac{1}{3}$ **17.** $5\frac{5}{9}$ **19.** $2\frac{4}{5}$ **21.** 7 paving stones **23.** 230.594 **25.** 45.7 **27.** $4\frac{2}{3}$ **29.** $10\frac{2}{3}$ **31.** $16\frac{1}{2}$

Block 5 Review

1. $\frac{1}{2} \times \frac{3}{4} = \frac{3}{8}$ **3.** $\frac{3}{5} \times \frac{3}{4} = \frac{9}{20}$ **5.** $\frac{2}{10} = \frac{1}{5}$ **7.** $\frac{1}{4}$ **9.** $\frac{5}{9}$ **11.** $\frac{3}{10}$ **13.** $\frac{1}{8}$ of the cars **15.** $\frac{1}{12}$ of the basket of chicken **17.** $\frac{3}{4} \div \frac{1}{4} = 3$
19. 3 **21.** 2 **23.** 4 **25.** $\frac{6}{7}$ **27.** $\frac{9}{10}$ **29.** 4 times **31.** 10 or 11 **33.** 2 **35.** 9 **37.** About 9 laps **39.** 8 **41.** 12 **43.** $25\frac{1}{2}$ **45.** 40 **47.** $4\frac{4}{5}$ **49.** $3\frac{5}{12}$ **51.** $2\frac{1}{10}$ **53.** $5\frac{5}{9}$ **55.** 6 days

BLOCK 6

Lesson 6.1

1. 10 units² **3.** $8\frac{7}{16}$ units² **5.** $\frac{11}{32}$ inch² **7.** $381\frac{1}{4}$ feet² **9.** $5\frac{4}{9}$ units² **11.** $14\frac{1}{16}$ inches² **13.** $10\frac{9}{16}$ feet² **15.** $1\frac{17}{64}$ inches² **17.** $\frac{1}{4}$ inch² **19.** $\frac{7}{8}$ inch² **21.** $3\frac{1}{16}$ inches² **23.** Answers may vary. **25.** $2\frac{7}{9}$ **27.** 11 **29.** $7\frac{3}{5}$

Lesson 6.2

1. P = 13.8 *cm* A = 7.7 *cm²* **3.** P = 14.4 *m* A = 7.475 *mm²* **5.** P = 17.6 *mm* A = 14.95 *mm²* **7. a)** 17.2 *cm* **b)** 18.49 *cm²* **9.** 6.8 *cm* **11.** 10.6 *cm* **13. a)** 95.2 *m* **b)** 566.44 *m²* **15.** 4.2 *cm²* **17.** 1.89 *cm²* **19.** $3\frac{1}{2}$ **21.** $\frac{1}{3}$ **23.** $5\frac{1}{4}$ **25.** 12 **27.** 9 portions

Lesson 6.3

1. 30 cm^2 **3.** 55 in^2 **5.** $14\frac{7}{8}$ in^2 **7.** See student work. 12 cm^2

9. See student work; 36 in^2

11. Answers may vary.

13. (1 rectangle and 2 triangles)

15. 153 yd^2 **17.** $21\frac{1}{2}$ in^2 **19.** 74 cm^2 **21. a)** 18 meters
b) 324 m^2 **23. a)** 32 ft^2 **b)** 4 ft^2 and 16 ft^2 **c)** One flower bed
gave more area. **25.** 10 boards **27.** No; he only has 99 cm of
cardboard. **29.** $79\frac{29}{32}$ in^2

Lesson 6.4

1. **3.** **5.**

7. a) **b)**

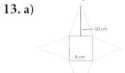 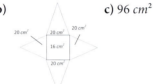

c) 158 m^2 **9.** 82 ft^2 **11.** 987 units^2
13. a) **b)** **c)** 96 cm^2

15. 216 in^2 **17. a)** 20 cm^2 **b)** 480 cm^2 **c)** 5760 cm^2 **d)** About
15 cans of paint **19.** $1\frac{1}{2}$ **21.** $4\frac{1}{10}$ **23.** $5\frac{2}{3}$

Lesson 6.5

1. $\frac{1}{8}$ cubic yard **3.** 28 cubic feet **5.** 20 cubic inches
7. a) $281\frac{2}{3}$ cubic yards **b)** no **9.** $4\frac{1}{2}$ cubic inches
11. $17\frac{1}{2}$ cubic feet **13.** $\frac{1}{27}$ cubic yard **15. a)** $\frac{8}{27}$ cubic inch
b) No. The volume of Ivan's cube will be 8 times greater than
Jenna's cube ($2\frac{10}{27}$ cubic inches). **17.** $7\frac{37}{100}$ cubic meters
19. 412.425 cubic inches or $412\frac{17}{40}$ cubic inches **21.** $\frac{5}{18}$ **23.** $\frac{4}{13}$
25. $1\frac{1}{2}$ **27.** $5\frac{1}{16}$ **29.** $3\frac{7}{16}$ in^2

Block 6 Review

1. $10\frac{1}{8}$ in^2 **3.** $6\frac{3}{4}$ in^2 **5.** $27\frac{1}{8}$ units^2 **7.** 4 in^2 **9.** $1\frac{1}{2}$ in^2
11. P = 30 m; A = 56.25 m^2 **13.** P = 8.7 cm; A = 2.96 cm^2
15. P = 15.6 cm; A = 15.12 cm^2 **17.** □ + △ + △ + △ + △
19. 85 m^2 **21.** 6 in^2 **23.** **25.** 39 cm^2 **27.** $\frac{27}{32}$ in^3
29. $38\frac{1}{2}$ yd^3
31. $3\frac{1}{9}$ yd^3 **33.** $29\frac{3}{4}$ ft^3

INDEX

A

Addition
  of decimals, 22
  of fractions, 113
    by renaming, 122
  of mixed numbers, 118
    by renaming, 122

Algorithm, standard
  of division, 45
  of multiplication, 35

Area
  area formulas, 169-170
  of composite figures, 178
  of rectangles, 169
  of squares, 169
  of triangles, 169
  using customary measurements, 169
  using metric measurements, 174

B

Base
  of a triangle, 170
  of a solid, 182

Base-ten blocks, 3, 44, 54

C

Career Focus
  Deputy Sheriff, 133
  Escrow Officer, 31
  Farmer, 66
  Metalsmith, 105
  Middle School Teacher, 198
  Reporter, 166

Centimeter, 11

Comparing
  decimals, 15
  fractions, 87
  symbols for, 16

Compatible numbers, 20, 151
  with decimals, 20
  with fractions, 152
  with mixed numbers, 152

Composite figures, 178

Composite number, 69

Cone, 182

Customary system
  unit of lengths
    foot, 96
    inch, 96
    Explore! Using a Customary Ruler, 96

Cylinder, 182

D

Decimal point, 3

Decimals, 3
  adding, 22
    Explore! Fit Occupations, 23
  base-ten blocks, 3, 44, 54
  comparing, 15
    Explore! Batting Averages, 16
  compatible numbers, 20
  difference of, 22
  dividing by decimals, 58
  dividing by whole numbers, 55
  equivalent, 15
  estimating with, 19
  measurement in centimeters, 11
  modeling, 3
  multiplying, 38
    Explore! Smart Shopper, 40
  ordering, 16
    Explore! Batting Averages, 16
  place value, 5
    Explore! Base-Ten Blocks, 5
  product of, 38
  quotient of, 54, 58
  reading and writing, 5
  rounding, 8
  subtracting, 22
    Explore! Fit Occupations, 23
  sum of, 22
  word form, 5

Denominator
  common, 69
  least common (LCD), 69
  unlike, 113

Dividend, 43

Division
  1-digit divisors, 43
    Explore! Beaded Necklaces, 44
  2-digit divisors, 49
    Explore! Magazine Subscriptions, 49
  dividend, 43
  divisor, 43
  multi-digit dividend, 43
  of decimals, 54, 58
  of fractions, 143, 147

Division
  of mixed numbers, 159
  of whole numbers, 43, 49
  quotient, 43
  reciprocals, 147
  remainder, 45
  standard algorithm, 45

Divisor, 43

E

Equivalent decimals, 15

Equivalent fractions, 74

Estimation
  decimals
    addition, 20
    compatible numbers, 20
    difference, 19
    division, 20
    multiplication, 19
    products, 19
    quotients, 20
    rounding, 19
    subtraction, 19
    sum, 20
  fractions
    addition, 108
    compatible numbers, 151
      Explore! 4-H Club, 151
    difference, 108
    division, 143, 147
    multiplication, 136, 139
    products, 136, 139
    quotients, 143, 147
    rounding, 108
    subtraction, 108
    sum, 108
  mixed numbers, 109
  symbols for, 16

Explore!
  4-H Club, 151
  Base-Ten Blocks, 5
  Batting Averages, 16
  Beaded Necklaces, 44
  Chocolate Chip Cookies, 93
  Creating Equivalent Fractions, 75
  Fit Occupations, 23
  Fraction Action, 136
  Fraction Homework, 79
  Magazine Subscriptions, 49
  Measuring Volume, 190
  Mixing Paint, 117
  Netting a Solid, 184
  Pizza Party, 112

Explore!
    Scrapbooking, 159
    Smart Shopper, 40
    Triangle Area, 170
    University Sales, 69
    Using a Customary Ruler, 96
    Using a Metric Ruler, 12
    What Fits?, 144
    Which is Larger?, 87

F
Faces, 182

Factor, 34
    common, 69
    greatest common (GCF), 69
    multiplication of, 34

Factorization, prime, 70, 84

Figures
    cylinder, 182
    cone, 182
    polygon, 182
    prism, 182
    pyramid, 182
    sphere, 182

Foot, 96

Fraction tiles, 74

Fraction, 45, 74
    adding, 113
        by renaming, 122
        Explore! Pizza Party, 112
    area using, 169
    comparing fractions, 87
        Explore! Which is Larger?, 87
    compatible numbers, 151
    denominator, 45, 74
    difference of, 113
        by renaming, 122
    dividing, 143, 147
        Explore! What Fits?, 144
    equivalent, 74
        Explore! Creating Equivalent
            Fractions, 75
    estimating with, 152
        Explore! 4-H Club, 151
    fraction tiles, 74
    improper, 92
        Explore! Chocolate Chip Cookies, 93
    measurement in inches, 96
    modeling, 74
    multiplying, 136, 139
        Explore! Fraction Action, 136
    numerator, 45, 47
    ordering, 87

Fraction
    Explore! Which is Larger?, 87
    proper, 92
    product of, 136, 139
    quotient of, 143, 147
    reciprocal of, 147
    renaming, 122
    rounding, 108
    simplest form, 79
        Explore! Fraction Homework, 79
    simplifying, 79
    subtracting, 113
        by renaming, 122
    sum of, 113
        by renaming, 122
    using models, 74
    volume using, 169
    whole numbers as, 92
    See Mixed numbers

G
Geometry
    area, see Area
    composite figure, see Composite figure
    cone, see Cone
    cylinder, see Cylinder
    perimeter, see Perimeter
    polygons, see Polygon
    prism, see Prism
    rectangles, see Rectangle
    rectangular prism, see Rectangular
        Prism
    solid, see Solid
    sphere, see Sphere
    square, see Square
    triangle, see Triangle

Greatest common factor, 69
    list, 69
    prime factorization, 70
    Venn diagram, 69
        Explore! University Sales, 69
    simplest form, 80

H
Height of a triangle, 170

I
Inch, 96

Improper fraction, 92
    writing as a mixed number, 92
        Explore! Chocolate Chip Cookies, 93

J

K
Kilometer, 11

L
Least common denominator (LCD), 85

Least common multiple (LCM), 83
    list, 83
    prime factorization, 84

Length
    customary system
        foot, 96
        inch, 96
    Metric system
        centimeter, 11
        kilometer, 11
        meter, 11
        millimeter, 11

Lowest terms, see Simplest form

M
Measurement
    area, 169, 174
    customary, 96
    metric, 11
    perimeter, 126, 174
    volume, 189

Meter, 11

Metric system
    units of length
        centimeter, 11
        Explore! Using a Metric Ruler, 12
        kilometer, 11
        meter, 11
        millimeter, 11

Millimeter, 11

Mixed numbers, 92
    adding, 118
        by renaming, 122
        Explore! Mixing Paint, 117
    difference of, 118
        by renaming, 122
    dividing, 159
        Explore! Scrapbooking, 159
    estimating, 109
    multiplying, 159
        Explore! Scrapbooking, 159
    product of, 159
    quotient of, 159
    renaming, 122
    rounding, 109
    subtracting, 118
        by renaming, 122
        Explore! Mixing Paint, 117

Mixed numbers
  sum of, 118
    by renaming, 122
  writing as improper fractions, 93
    Explore! Chocolate Chip Cookies, 93

Models
  base-ten blocks, 3
  dividing decimals, 54, 58
  dividing fractions, 143
    Explore! What Fits?, 144
  fraction tiles, 74
  multiplying decimals, 38
    Explore! Smart Shopper, 40
  multiplying fractions, 136, 139
    Explore! Fraction Action, 136

Multiples, 83
  least common (LCM), 83

Multiplication
  factor, 34
  multi-digit factors, 34
  of decimals, 38
  of fractions, 136, 139
  of mixed numbers, 159
  product, 34
  standard algorithm, 35

N
Net, 183
  Explore! Netting a Solid, 184
  surface area using, 184

Numbers
  compatible, 20
  composite, 69
  mixed, 92
  prime, 69

Numerator, 45, 74

O
Ordering
  decimals, 15
  fractions, 87

P
Perimeter
  using customary measurements, 126
  using metric measurements, 174

Perpendicular, 170

Place value, 3
  Explore! Base-Ten Blocks, 5

Point, decimal, 3

Polygon, 126
  area, 169
  perimeter, 126

Prime factorization, 70
  Greatest common factor, 70
  Least common multiple, 84

Prime number, 69

Prism, 182

Product
  of decimals, 38
  of factors, 34
  of fractions, 136, 139
  of mixed numbers, 159
  of whole numbers, 34

Proper fraction, 92

Pyramid, 182

Q
Quotient
  of decimals, 55, 58
  of fractions, 143, 147
  of mixed numbers, 159
  of whole numbers, 43, 49

R
Reciprocals, 147

Rectangle
  area of, 169
  area formula, 169
  perimeter, 126

Rectangular prism, 182, 189

Reduced fraction, see Simplest form

Remainder, 45

Rounding
  decimals, 8
  fractions, 108
  mixed numbers, 109

S
Simplest form, 79
  common factors, 80
  greatest common factor, 80

Solid, 182

Sphere, 182

Square
  area formula, 170
  area of, 170

Standard algorithm
  of division, 45
  of multiplication, 35

Subtraction
  of decimals, 22
  of fractions, 113
    by renaming, 122
  of mixed numbers, 118
    by renaming, 122

Surface area, 184

T
Tick marks, 11

Triangle
  area of, 170
    Explore! Triangle Area, 170
  area formula, 170
  base of, 170
  height of, 170
  vertex of, 170

U
Units, see Customary system and Metric System

V
Venn diagram, 69

Volume, 189
  Explore! Measuring Volume, 190
  formula for
    rectangular prism, 189
    for prism, 192

W
Whole numbers
  as fractions, 92
  multiplying by fractions, 155
  dividing decimals by, 55
  dividing by fractions, 155

X

Y

Z

PROBLEM-SOLVING

UNDERSTAND THE SITUATION

- ► Read then re-read the problem.
- ► Identify what the problem is asking you to find.
- ► Locate the key information.

PLAN YOUR APPROACH

Choose a strategy to solve the problem:

- ► Guess, check and revise
- ► Use an equation
- ► Use a formula
- ► Draw a picture
- ► Draw a graph
- ► Make a table
- ► Make a chart
- ► Make a list
- ► Look for patterns
- ► Compute or simplify

STOP AND THINK

- ► Did you answer the question that was asked?
- ► Does your answer make sense?
- ► Does your answer have the correct units?
- ► Look back over your work and correct any mistakes.

SOLVE THE PROBLEM

- ► Use your strategy to solve the problem.
- ► Show all work.

DEFEND YOUR ANSWER

Show that your answer is correct by doing one of the following:

- ► Use a second strategy to get the same answer.
- ► Verify that your first calculations are accurate by repeating your process.

ANSWER THE QUESTION

- ► State your answer in a complete sentence.
- ► Include the appropriate units.

SYMBOLS

Algebra and Number Operations

SYMBOL	MEANING
+	Plus or positive
−	Minus or negative
$5 \times n$, $5 \cdot n$, $5n$, $5(n)$	Times (multiplication)
$3 \div 4$, $4\overline{)3}$, $\frac{3}{4}$	Divided by (division)
=	Is equal to
≈	Is approximately
<	Is less than
>	Is greater than
%	Percent
$a : b$ or $\frac{a}{b}$	Ratio of a to b
$5.\overline{2}$	Repeating decimal (5.222...)
≥	Is greater than or equal to
≤	Is less than or equal to
x^n	The nth power of x
(a, b)	Ordered pair where a is the x-coordinate and b is the y-coordinate
±	Plus or minus
$\sqrt{x}$	Square root of x
≠	Not equal to
$x \overset{?}{=} y$	Is x equal to y?
$\lvert x \rvert$	Absolute value of x
P(A)	Probability of event A

Geometry and Measurement

SYMBOL	MEANING
≅	Is congruent to
~	Is similar to
∠	Angle
$m\angle$	Measure of angle
$\triangle ABC$	Triangle ABC
$\overline{AB}$	Line segment AB
$\overrightarrow{AB}$	Ray AB
AB	Length of AB
π	Pi (approximately $\frac{22}{7}$ or 3.14)
°	Degree